计算机应用专业

实用计算机英语

Shiyong Jisuanji Yingyu

（第4版）

主　编　於　芳

副主编　钱湘莺　吴志燕　任吉丹
　　　　高森凤　杜婷婷

主　审　于丽娟

中国教育出版传媒集团

高等教育出版社·北京

内容提要

本书是"十四五"职业教育国家规划教材，在第3版的基础上修订而成。本书根据中等职业学校学生的实际情况，结合计算机技术及其应用的最新发展状况，对软件版本与硬件知识进行了更新或升级，以求与时俱进，从而给学生更大的帮助。本书较好地解决了在计算机专业知识学习过程中，如何进行英语语言学习、培养语言技能的问题。

本书的语言素材取自日常的计算机使用场景，内容与人们的工作和生活密切相关，包括计算机的基础知识、硬件、软件、操作系统、办公自动化、Internet、多媒体、创建网站、网络安全、计算机维护等。

扫描书中的二维码，可随时收听重点词汇和课文录音。

本书配套电子课件、教参等辅教辅学资源，请登录高等教育出版社Abook新形态教材网（http://abook.hep.com.cn）获取相关资源。详细使用方法见本书最后一页"郑重声明"下方的"学习卡账号使用说明"。

本书适合中职计算机类专业学生使用，也可供广大技术人员及计算机爱好者阅读。

图书在版编目（CIP）数据

实用计算机英语 / 於芳主编. -- 4版. -- 北京：高等教育出版社，2023.8

计算机应用专业

ISBN 978-7-04-060794-9

Ⅰ.①实… Ⅱ.①於… Ⅲ.①电子计算机-英语-中等专业学校-教材　Ⅳ.①TP3

中国国家版本馆CIP数据核字(2023)第123153号

策划编辑	郭福生	责任编辑	周海燕	封面设计	李小璐	版式设计	于 婕
责任绘图	邓 超	责任校对	刁丽丽	责任印制	存 怡		

出版发行	高等教育出版社	网　址	http://www.hep.edu.cn	
社　址	北京市西城区德外大街4号		http://www.hep.com.cn	
邮政编码	100120	网上订购	http://www.hepmall.com.cn	
印　刷	北京市密东印刷有限公司		http://www.hepmall.com	
开　本	889 mm×1194 mm　1/16		http://www.hepmall.cn	
印　张	8.75	版　次	2007年7月第1版	
字　数	180千字		2023年8月第4版	
购书热线	010-58581118	印　次	2023年8月第1次印刷	
咨询电话	400-810-0598	定　价	38.20元	

本书如有缺页、倒页、脱页等质量问题，请到所购图书销售部门联系调换

版权所有　侵权必究

物　料　号　60794-00

前　言

本书第 1 版于 2007 年出版，并于 2013 年、2017 年进行了两次修订，在全国各地的中等职业学校和技工院校得到了广泛使用，其知识性、实用性和趣味性深受师生们的喜爱，也颇受计算机英语爱好者的欢迎。在本次修订过程中，依然秉承"中职英语教学为专业学习服务、为学会生活服务、为科技进步服务"的理念，以计算机知识为主线、以英语为载体进行编写，与时俱进地反映了当前数字经济蓬勃发展、信息技术普遍应用于工作与生活各领域的情况，对许多知识内容和图片进行了更新，着重体现了我国科技自主创新的成果，并有机融入课程思政元素，以学习贯彻党的二十大精神，落实立德树人根本任务。

本书共 10 个单元，每个单元都由 4 篇课文组成，第一课是导入内容，学习与单元主题相关的计算机知识和英语词汇，为后面的学习做好铺垫；第二课是阅读课文及相关的理解练习，进一步深入学习相关计算机和语言知识；第三课以听说训练为主，围绕相关热点现象展开交流讨论，进行语言输出训练；第四课以练习为主，对所学的知识和词汇进行复习巩固。本书的内容除了介绍常用的计算机组成、外围设备、操作系统、应用软件等内容外，还包括计算机维护、计算机常用英语等工作中十分常用的内容，有较强的实用性和可操作性，使用本书进行学习，不仅可以掌握常用的计算机专业英语词汇，还可以学习或巩固基本的计算机操作技能。

本书的主要特色有五个方面：一是扫码可收听课文录音，并学习相关课外知识，使用很方便；二是选材丰富、贴近生活，并且符合当下数字经济时代的技术发展趋势；三是较好地结合了英语与计算机两门学科的知识与技能，具有较强的实用性和可操作性；四是学习内容深入浅出，表现形式通俗易懂，符合中职学生的学习水平和身心特点；五是每个单元均以拓展阅读的形式给学生提供阅读材料，弘扬社会主义核心价值观，加强科技自立自强的信念教育。

本书配套电子课件、教参等辅教辅学资源，请登录高等教育出版社 Abook 新形态教材网（http://abook.hep.com.cn）获取相关资源。详细使用方法见本书最后一页"郑重声明"下方的

"学习卡账号使用说明"。

本书由浙江省绍兴市中等专业学校於芳任主编,并负责全书内容的整体规划和统稿,各单元编写分工如下:钱湘莺编写 Unit 1~2,吴志燕编写 Unit 3~4,任吉丹编写 Unit 5~6,高森凤编写 Unit 7~8,杜婷婷编写 Unit 9~10。本书由浙江省教育科学研究院职业与成人教育研究所副所长于丽娟担任主审,编写也得到了绍兴文理学院罗军平、绍兴市中等专业学校王永昌和沈银燕等计算机专业教师的支持,朗读录音由绍兴文理学院在读的英语专业大学生团队录制,在此一并表示感谢。

希望此书能够为中职学生学习计算机知识和英语知识提供一定的帮助。由于编者水平有限,不足之处在所难免,恳请广大读者不吝批评指正并提出宝贵意见。读者意见反馈邮箱:zz_dzyj@pub.hep.cn。

编　者

2023 年 6 月

Contents

Unit 1 Brief Introduction to Computer   1

    Lesson One ················· 1
    Lesson Two ················· 2
    Lesson Three ··············· 6
    Lesson Four ················ 7

Unit 2 Computer Hardware ············ 10

    Lesson Five ················ 10
    Lesson Six ················· 12
    Lesson Seven ··············· 15
    Lesson Eight ··············· 17

Unit 3 Computer Software ············ 21

    Lesson Nine ················ 21
    Lesson Ten ················· 22
    Lesson Eleven ·············· 25
    Lesson Twelve ·············· 27

Unit 4 Operating System ············· 31

    Lesson Thirteen ············ 31
    Lesson Fourteen ············ 33

    Lesson Fifteen ············· 36
    Lesson Sixteen ············· 38

Unit 5 Office Automation ············ 42

    Lesson Seventeen ··········· 42
    Lesson Eighteen ············ 45
    Lesson Nineteen ············ 48
    Lesson Twenty ·············· 53

Unit 6 The Internet ················· 58

    Lesson Twenty-one ·········· 58
    Lesson Twenty-two ·········· 59
    Lesson Twenty-three ········ 63
    Lesson Twenty-four ········· 65

Unit 7 Multimedia ··················· 70

    Lesson Twenty-five ········· 70
    Lesson Twenty-six ·········· 72
    Lesson Twenty-seven ········ 76
    Lesson Twenty-eight ········ 78

Unit 8　Creating a Website ·················82

　　　　Lesson Twenty-nine ···············82
　　　　Lesson Thirty ·····················84
　　　　Lesson Thirty-one ···············87
　　　　Lesson Thirty-two ···············90

Unit 9　Network Security ····················98

　　　　Lesson Thirty-three ············98
　　　　Lesson Thirty-four ············100

　　　　Lesson Thirty-five ············103
　　　　Lesson Thirty-six ············106

Unit 10　Computer Maintenance ······110

　　　　Lesson Thirty-seven ········110
　　　　Lesson Thirty-eight ········112
　　　　Lesson Thirty-nine ············115
　　　　Lesson Forty ·····················117

Glossary··121

Unit 1 Brief Introduction to Computer

Lesson One

○ **Match the pictures with their English names.**

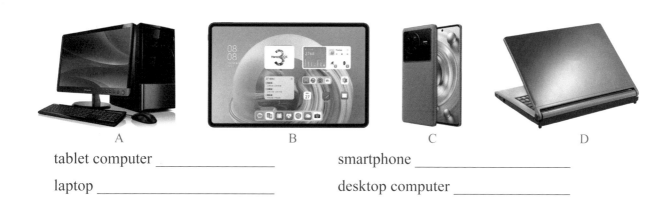

A B C D

tablet computer _____ smartphone _____

laptop _____ desktop computer _____

○ **Listen and fill in the blanks with the proper forms of the words or phrases given.**

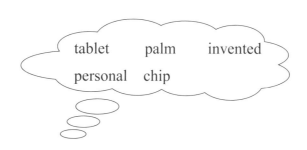

tablet   palm   invented
personal   chip

1. Computer is _____（被发明）for performing numerical calculations（进行数值计算）.

2. A _____ (芯片) is a kind of microprocessor (微处理器) in computers.

3. _____ (个人的) computer is designed for one person to use at work or at home.

4. _____ (平板) computers are becoming more and more popular.

5. The smartphone is a complex of _____ (手掌) computer and mobile phone.

Word Bank

tablet /ˈtæblət/ n. 平板电脑
desktop /ˈdesktɒp/ adj. 台式的
invent /ɪnˈvent/ v. 发明；创造
chip /tʃɪp/ n. 芯片
launch /lɔːntʃ/ v. 推出（新产品）
complex /ˈkɒmpleks/ n. 复合体
smartphone /ˈsmɑːtfəʊn/ n. 智能手机

smart /smɑːt/ adj. 聪明的，智能的
laptop /ˈlæptɒp/ n. 笔记本电脑
personal /ˈpɜːsən(ə)l/ adj. 个人的
corporation /ˌkɔːpəˈreɪʃ(ə)n/ n. 公司
microprocessor /ˌmaɪkrəʊˈprəʊsesə(r)/ n. 微处理器
palm /pɑːm/ n. 手掌

Lesson Two

- What changes have computers brought to our life?
- What will the future computer be like?

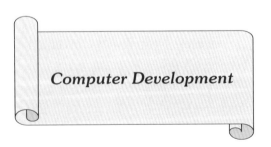

A computer is a fast data operating system to accept, store, process data and produce output

ENIAC

results. The first computer was invented by John Mauchly and J. Presper Eckert in 1946, and it was known as ENIAC.

The modern computers have come through four generations: vacuum tube computers, transistor computers, small-scale integrated circuit computers, and large-scale or super large-scale integrated circuit computers. CDC 1604 (the first supercomputer in the world, designed by Seymour Cray) is a typical example. Nowadays, more and more supercomputing facilities have been applied to study weather, biomedicine, new energy, and many other fields.

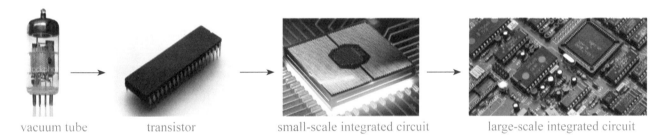

vacuum tube   transistor   small-scale integrated circuit   large-scale integrated circuit

Today we are facing a computer revolution. The computers in the future will become giant, intelligent, miniature, networking and multimedia. The technologies of Bluetooth, Wi-Fi, Cloud Computing, the Internet of Things and the Intelligent Earth are bringing great changes to our life.

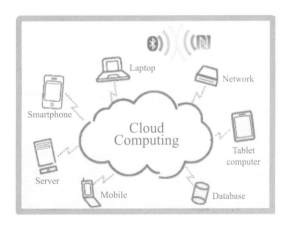

Word Bank

data /'deɪtə/ n. 数据；资料
process /'prəʊses/ v. 加工；处理
generation /ˌdʒenə'reɪʃ(ə)n/ n. 一代
tube /tjuːb/ n. 管子
scale /skeɪl/ n. 规模

circuit /'sɜːkɪt/ n. 电路
typical /'tɪpɪk(ə)l/ adj. 典型的
supercomputer /'sjuːpəkəmpjuːtə(r)/ n. 超级计算机
facility /fə'sɪləti/ n. 设施，设备
energy /'enədʒi/ n. 能源
revolution /ˌrevə'luːʃn/ n. 改革
intelligent /ɪn'telɪdʒənt/ adj. 智能的
networking /'netwɜːkɪŋ/ n. 网络化
technology /tek'nɒlədʒɪ/ n. 技术

operate /'ɒpəreɪt/ v. 操作
output /'aʊtpʊt/ n. 输出
vacuum /'vækjuːm/ adj. 真空的
transistor /træn'zɪstə(r)/ n. [电子] 晶体管
integrated /'ɪntɪɡreɪtɪd/ adj. 综合的，完整的，集成的

design /dɪ'zaɪn/ v. 设计
apply /ə'plaɪ/ v. 适用，应用

biomedicine /ˌbaɪəʊ'medɪsɪn/ n. 生物医学
field /fiːld/ n. 领域
giant /'dʒaɪənt/ adj. 巨大的
miniature /'mɪnɪtʃə(r)/ adj. 微型的
multimedia /ˌmʌltɪ'miːdiə/ n. 多媒体
Bluetooth 蓝牙

Wi-Fi (wireless fidelity) 基于 IEEE 802.11b 标准的无线局域网
Cloud Computing 云计算
the Intelligent Earth 智能地球
the Internet of Things 物联网

 Choose the best answers according to the text.

1. The first computer was called _____.
   A. John Mauchly   B. J. Presper Eckert
   C. ENIAC

2. The computers in the first generation were called _____.
   A. transistor computers   B. vacuum tube computers
   C. integrated circuit computers

3. The fourth generation computers were called _____.
   A. vacuum tube computers   B. integrated circuit computers
   C. large-scale or super large-scale integrated circuit computers

4. The future computers will become _____.

   A. intelligent and miniature
   B. networking and multimedia
   C. giant, intelligent, miniature, networking and multimedia

Match the pictures with the words.

miniature _____

multimedia _____

networking _____

intelligent _____

Fill in the blanks with the words according to the text.

| four | computers | vacuum | small-scale |
| modern | future | transistors | |

The _____ computers have experienced _____ generations. The computers in the first generation, which used vacuum tubes, were called _____ tube computers. In the second generation _____ replaced vacuum tubes, so the computers in that period of time were called transistor computers.

The computers followed in the 1970s by integrated circuits in the third generation were called _____ integrated circuit computers. The fourth generation were called large-scale or super large-scale integrated circuit _____.

The computers in the _____ will tend to become giant, intelligent, miniature, networking and multimedia.

Lesson Three

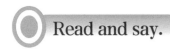 Read and say.

- What should you do before you buy a computer?
  You should know how to use the computer and how much you can afford.
- Where can you buy a computer?
  In a supermarket, at an online store or in an electronics shop.
- How much memory will you buy?
  It depends on the size of the software you plan to run.
- How much should you spend on your computer?
  As much as you can afford.
- Which kind of computer will you buy, a laptop or a desktop?
  A laptop is portable while a desktop has a larger screen and is expandable.

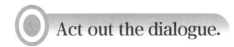 Act out the dialogue.

A Desktop Or A Laptop?

Tom: Hi, Li Lei. Long time no see. What's happening?
Li Lei: Everything is OK. What's going on?

Tom: Well, I want to buy a computer. Could you give me some ideas?

Li Lei: In general, a laptop is convenient to carry while a desktop is expandable and is often used in the office or at home.

Tom: Oh, I see. I only use the computer at home, so I'm going to buy a desktop computer. Thank you for your advice.

Li Lei: You're welcome.

Word Bank

electronics /ɪˌlek'trɒnɪks/ n. 电子器件
depend /dɪ'pend/ v. 依靠
afford /ə'fɔːd/ v. 负担得起
screen /skriːn/ n. 屏幕
depend on 取决于

memory /'meməri/ n. 存储器
size /saɪz/ n. 大小；规模
portable /'pɔːtəb(ə)l/ adj. 轻便的，便携式的
expandable /ɪk'spændəbl/ adj. 可扩展的
in general 一般而言；通常

Fill in the blanks and make a plan to buy a computer.

How much money to buy a computer	¥5,000
Where to use the computer	
What kind of computer to buy	
Why to choose this kind of computer	
How much memory to choose	
Where to buy the computer	

Lesson Four

Translate the following abbreviations into Chinese according to the pictures.

CPU_____

DDR_____

CD-R_____

  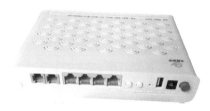

GPU_____   SSD_____   Modem_____

Put the steps in the right order according to the question.

1. How can you start a program?

(   ) A. Click any item（项目）on the Start menu without an arrow or ellipsis（省略号）to start that program.

(   ) B. Click the Start button to reveal（显示）the Start menu.

2. How can you shut down a computer properly?

(   ) A. Cut off the electricity（电）.

(   ) B. Click the Start button to find the item "Shut down the computer".

(   ) C. Make sure you have withdrawn（退出）from all programs（程序）.

Tick out (√) the advantages of computers.

☐ 1. It is very convenient to correct（改正）your spelling mistakes（错误）on the computer.

☐ 2. We can store a lot of information（信息）on the computer.

☐ 3. It is bad for your health（健康）to stay in front of the computer for a long time.

☐ 4. Computers can make teaching and learning more vivid（生动的）and interactive（交互的）.

☐ 5. People who use computers too much may be over-dependent on computers.

☐ 6. Business can benefit（获益）from the wide application（应用）of computers.

Match the following positions（岗位）with their Chinese names.

(   ) Applications Programmer   A. 数据库管理员

(　　) Software Engineer　　　　B. 软件工程师
(　　) Database Administrator　　C. 应用软件程序员
(　　) 3D Animator　　　　　　　D. 室内设计师
(　　) Web Editor　　　　　　　　E. 三维动画设计师
(　　) Interior Designer　　　　　F. 网络编辑

Further reading.
拓展阅读

China's Supercomputer

　　In 1983, China's first supercomputer was born. It made China be the third country in the world who had the ability of developing supercomputers. From "Galaxy"（银河超级计算机）to "Tianhe", "Sunway" series of supercomputers（天河、神威系列）, China's supercomputing sector（领域）is among global front-runners（全球领先地位）. The "Sunway TaihuLight" was built entirely using processors designed and manufactured in China（中国设计、中国制造的处理器）.

▼ After reading, how do you feel?
☐ proud　☐ upset　☐ inspired（受鼓舞的）　☐ other feelings: _____

▼ Discuss:
What do you learn from the passage above?

Hints: the importance of self-reliance（自立）and self-improvement（自强）in science and technology（科技）

Unit 2   Computer Hardware

Lesson Five

Match the pictures with the hardware names.

A — printer
B — webcam
C — USB disk
D — headphone
E — monitor
F — keyboard

webcam _____   USB disk _____   printer _____
monitor _____   headphone _____   keyboard _____

Tell the names of the following hardware, using the words given.

| memory | CPU | mainboard | hard disk |
| display card | fan | sound card | power supply |

1. _____ 2. _____ 3. _____ 4. _____

5. _____ 6. _____ 7. _____ 8. _____

 Listen and fill in the blanks.

```
send      print      store
display   type       device
```

- A webcam is a video camera used to _____（发送）and display electronic pictures over the Internet.
- A USB disk can easily _____（存储）and take data.
- A printer is a device to _____（打印）files.
- A monitor is a computer_____（显示器）on which words or pictures can be shown.
- Headphones are a pair of small _____（设备）through which you can listen to music from your computer.
- A computer keyboard is an input device from which you can _____（输入）text or commands.

Word Bank

video /ˈvɪdiəʊ/ n. 录像；视频
disk /dɪsk/ n. 磁盘
monitor /ˈmɒnɪtə(r)/ n. 显示器

camera /ˈkæm(ə)rə/ n. 照相机
printer /ˈprɪntə(r)/ n. 打印机
headphone /ˈhedfəʊn/ n. 耳机

keyboard /'ki:bɔ:d/ *n.* 键盘
display /dɪ'spleɪ/ *v.* 显示；*n.* 显示器
power /'paʊə(r)/ *n.* 电力
device /dɪ'vaɪs/ *n.* 装置；设备
send /send/ *v.* 发送
store /stɔ:(r)/ *v.* 存储

mainboard /'meɪnbɔ:d/ *n.* 主板
fan /fæn/ *n.* 风扇
supply /sə'plaɪ/ *n.* 供应
type /taɪp/ *v.* 打字；输入
command /kə'mɑ:nd/ *n.* 指令

Lesson Six

 Discussion.

1. What are the basic output devices of a computer?
2. How many kinds of computer hardware are there? And what are they?

 Reading.

 扫一扫 听录音

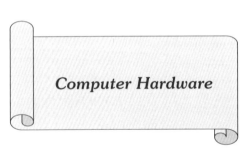

Computer Hardware

There are four kinds of computer hardware: CPU, storage devices, input devices and output devices.

● The central processing unit (CPU)

It is the heart of the computer. The design of the CPU affects the processing capacity and the speed of the computer. Multi-core processor is popular today.

● Storage devices

We usually divide the storage devices into two types: the main memory and the secondary storage. The main memory often refers to RAM (Random Access Memory) and ROM (Read-Only Memory). ROM can be read, but can't be written. USB disks and hard disks are two kinds of the secondary storage.

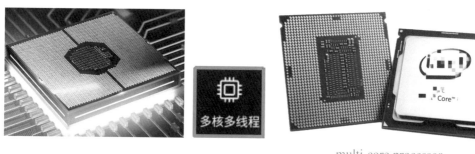

multi-core processor

USB disk

hard disk

- Input devices

The most common input devices are the keyboard and the mouse. The keyboard is used to give instructions to a computer. The mouse helps you select icons or items on your screen.

keyboard

mouse

- Output devices

The monitor and the video card work together to display texts, images and videos on the screen. The printer is also a kind of output device.

monitor

printer

Word Bank

storage /'stɔːrɪdʒ/ *n.* 存储器，存储
unit /'juːnɪt/ *n.* 部件；组件
capacity /kə'pæsəti/ *n.* 容量
secondary /'sekənd(ə)ri/ *adj.* 辅助的
access /'ækses/ *n.* 入口；访问，存取
instruction /ɪn'strʌkʃ(ə)n/ *n.* 指令
icon /'aɪkɒn/ *n.* 图标
divide ... into ... 把……分成……

central /'sentrəl/ *adj.* 中心的；主要的
affect /ə'fekt/ *v.* 影响
multi-core /'mʌltɪ'kɔː/ *adj.* 多核心的
random /'rændəm/ *adj.* 随机的
common /'kɒmən/ *adj.* 常见的
select /sɪ'lekt/ *v.* 选择；挑选
mouse /maʊs/ *n.* 鼠标［器］
refer to 指的是；提及

Decide whether the following statements are true (T) or false (F) according to the passage.

(　　) 1. CPU has nothing to do with the processing capacity.
(　　) 2. The heart of the computer is the main memory storage.
(　　) 3. The storage devices can be divided into two parts: the main memory and the secondary storage.
(　　) 4. ROM can be written, but can't be read.
(　　) 5. A video card can display texts and images.

Divide the following words into categories.

keyboard　　monitor　　CPU　　USB disk　　RAM　　mouse
ROM　　microphone　　printer　　hard disk　　headphone

Input Device(s)	Output Device(s)	Storage Device(s)	Processor Unit(s)

Complete the following sentences according to the passage.

| display | divided into | refers to | give instructions to | speed |

1. CPU affects the _____ of the computer.
2. The monitor and the video card work together to _____ texts, images and videos.
3. The keyboard lets you _____ a computer.
4. The storage devices can be _____ two types.
5. The main memory _____ ROM and RAM.

Lesson Seven

Read and say.

- What is the heart of a computer?
  CPU. It is a kind of chips, the "brain" of the electronic products.
- What is a "chip"?
  It is an integrated circuit（集成电路）.
- Where can chips be found?
  Chips lie inside a wide range of products, such as smartphones, computers, automobiles and other equipment. CPUs and GPUs are the most common chips in computers.
- How many types of Integrated Circuits (ICs)?
  3 types: digital ICs, analog ICs and Mixed-Signal ICs. CPU is just one kind of digital ICs.

Act out the dialogue.

(*Tom and Li Lei are talking about computer chips.*)

Tom: My PC (Personal Computer) doesn't work. Can you help me to check it?

Li Lei: Sure. Let me see. Well, something is wrong with the CPU. This kind of CPU has almost been eliminated. It's impossible to buy such a new one.

Tom: What a pity! How can we choose the CPU?

Li Lei: There are two giant CPU manufacturers, Intel and AMD. For example, Intel's CPU types Core-i5, Core-i7, even Core-i9, which has 18 cores.

Tom: Do supercomputers have CPUs?

Li Lei: Yes. Sunway TaihuLight, one of China's supercomputers, is using China's own chips named SW26010. It has 40,960 CPUs, all of which were designed in China.

Tom: Amazing! Does China have her own CPU manufacturers?

Li Lei: Yes. There are many famous China's CPU brands, like Kunpeng, Phytium, HYGON, LOONGSON, Zhaoxin, SUNWAY, and so on.

Tom: That's amazing. Thank you for your introduction.

Li Lei: You're welcome.

Answer the following questions according to the dialogue.

1. What are they talking about?

2. What are the famous CPU brands?

3. What CPU brands belong to China?

Word Bank

扫一扫
听录音

electronic /ɪˌlek'trɒnɪk/ adj. 电子的，电子学的
product /'prɒdʌkt/ n. 产品
automobile /'ɔːtəməbiːl/ n. 汽车
equipment /ɪ'kwɪpmənt/ n. 设备
GPU abbr. 图形处理器（Graphics Processing Unit）
analog /'ænəlɒg/ adj. 模拟的
mixed-signal /mɪkst'sɪgnəl/ adj. 混合信号的
eliminate /ɪ'lɪmɪneɪt/ v. 淘汰
digital ICs 数字芯片

manufacturer /ˌmænju'fæktʃərə(r)/ n. 制造商
core /kɔː(r)/ n. 核心
a wide range of ...　大范围的，各种不同的
Intel（美国）英特尔公司
AMD (Advanced Micro Devices, Inc.)（美国）超威半导体公司
Sunway TaihuLight　神威·太湖之光（中国超级计算机）
Zhaoxin 兆芯

Mixed-Signal ICs 混合信号芯片
Phytium 飞腾
LOONGSON 龙芯
SUNWAY 神威
Kunpeng 鲲鹏

SW26010　申威 26010（一款中国自主设计、制造的高性能多线程处理器）
analog ICs　模拟芯片
HYGON 海光

Lesson Eight

Put the following sentences into the right order. Learn how to insert and remove the removable storage devices.

1. Insert the flash drive（闪存盘）into the USB port（接口）and see whether the flash drive appears（出现）.

2. Copy the files（文件）or folders（文件夹）from the flash drive to the computer.

3. Click **My Computer**. If you have inserted your flash drive properly（正确地）, you can find it in the **My Computer** window.

4. When the copying is finished, do not immediately（立即）remove（移除）the flash drive from the USB port.

5. Click on the **Safely Remove Mass Storage Device**. When you see **Safe To Remove Hardware**, remove your flash drive.

Computer Hardware Checklist.

Before you walk into a computer store to buy a desktop computer, go over the list of hardware that you might need. Please put the following list into Chinese and select the necessary parts to assemble（组装）a computer.

Computer Hardware Checklist

CPU _____ (   )

Mainboard _____ (   )

CD-ROM _____ (   )

Monitor _____ (   )

Video Card _____ (   )

Chassis _____ (   )

Keyboard _____ (   )

Mouse _____ (   )

Printer _____ (   )

Fax _____ (   )

Modem _____ (   )

Sound Card _____ (   )

Memory _____ (   )

Hard Disk _____ (   )

Connect the following computer external devices（外部设备）to the right ports on the chassis.

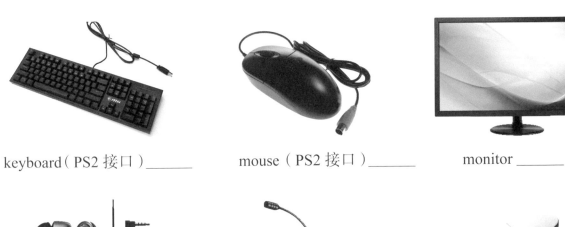

keyboard（PS2 接口）_____   mouse（PS2 接口）_____   monitor _____

earphones _____

microphone _____

USB disk _____

dot matrix printer _____   network wire _____   power line _____

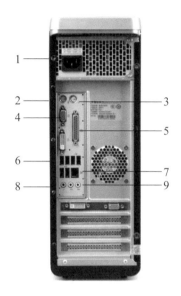

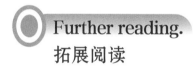

Further reading.
拓展阅读

The Brain-Computer Interface Chips

China released its first BCI (Brain-Computer Interface 脑机接口技术) chip, named "Brain Talker". It can read brain activity. The chip was released (发布) at The 3rd World Intelligence Congress (世界智能大会) held in Tianjin. It was jointly developed (联合研发) by Tianjin University and China Electronics Cooperation ("CEC" 中国电子信息产业集团有限公司). China holds the fully independent intellectual property (拥有完全自主知识产权).

The chip can be used to medical treatment, e.g. brain-typing —— using BCI technology, some motor-impaired people（运动障碍的人）could type by their brains to "talk".

▼ **After reading, are you proud of our country? Where can chips be applied?**

(Note: be applied to 应用于……领域)

▼ **Survey:**

Why does China attach great importance to the independent chip research and development? (Note: attach great importance to... 十分重视)

Hints: technology bottleneck 技术瓶颈，key and core technologies 关键核心技术，national strategy 国家战略，industry chain 产业链

Unit 3 Computer Software

Lesson Nine

○ Match the icons with the software names.

Maya   Windows   CorelDRAW   360 Antivirus   Photoshop   WPS

○ Find the right function for the software.

Software   *Functions*

    Play music and videos.

    Kill viruses.

    Process word files.

    Produce animation.

    Chat online.

Edit images and graphics.

Listen and fill in the blanks.

扫一扫 听录音

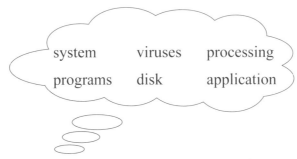

- Computer software is a set of computer _____（程序）.
- There are two kinds of software. They are _____（系统）software and application software.
- System software provides support for _____（应用）software.
- With system software, you can write information to a _____（磁盘）, check for _____（病毒）and so on.
- Application software performs useful work such as word and image _____（处理）.

扫一扫 听录音

Word Bank

graphic /ˈɡræfɪk/ *n.* 图形，图表
system /ˈsɪstəm/ *n.* 系统
support /səˈpɔːt/ *v.* 支持
perform /pəˈfɔːm/ *v.* 执行，履行

animation /ˌænɪˈmeɪʃ(ə)n/ *n.* 动画
application /ˌæplɪˈkeɪʃ(ə)n/ *n.* 应用
virus /ˈvaɪrəs/ *n.* 病毒

Lesson Ten

 Discussion.

1. There are two kinds of software. What are they?
2. Do they have special functions?

Reading.

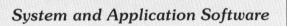

System and Application Software

System software is an important part of computer systems. It is used to control computer management and to maintain computer software and hardware resources. System software mainly includes the following types:

- Operating system software, such as Mac OS, Windows, UNIX, Linux, etc.
- Language processing systems, such as C sharp, Java, Visual Basic, etc.
- DBMSs (database management systems), such as Oracle, MySQL, SQL Server, DB2, etc.

Application software can help you to do specific tasks. Here are several common kinds:

- Word processors, including Microsoft Word, WPS Office, Notepad, etc.
- Image processing software, including Photoshop, CorelDRAW, etc.
- Computer aided design software, including AutoCAD, CorelCAD, etc.
- Input method editors, including Sogou Pinyin Input Method, Five-stroke Input Method, Baidu Pinyin Input Method, etc.
- Translation tools, such as Kingsoft PowerWord, Kingsoft FastAIT, Youdao Dict, etc.
- Browsers, including Edge, Maxthon, Firefox, Sogou Explorer, etc.
- Electronic reading tools, such as Adobe Reader, Foxit Reader, etc.

Word Bank

special /ˈspeʃ(ə)l/ *adj.* 特殊的
control /kənˈtrəʊl/ *v.* 控制，管理
maintain /meɪnˈteɪn/ *v.* 维持
include /ɪnˈkluːd/ *v.* 包含，包括
task /tɑːsk/ *n.* 任务

function /ˈfʌŋkʃ(ə)n/ *n.* 功能
management /ˈmænɪdʒmənt/ *n.* 管理
resource /rɪˈsɔːs/ *n.* 资源
specific /spɪˈsɪfɪk/ *adj.* 明确的；特殊的
processor /ˈprəʊsesə(r)/ *n.* 处理器；处理程序

including /ɪnˈkluːdɪŋ/ prep. 包括……在内
aid /eɪd/ n. 帮助
translation /trænzˈleɪʃ(ə)n/ n. 翻译
browser /ˈbraʊzə(r)/ n. 浏览器
design /dɪˈzaɪn/ v. 设计
method /ˈmeθəd/ n. 方法
tool /tuːl/ n. 工具
electronic /ɪˌlekˈtrɒnɪk/ adj. 电子的

DBMS: Database Management System 数据库管理系统
CAD: Computer-Aided Design 计算机辅助设计

Write out the software types.

AutoCAD, CorelCAD　　　　　　＿＿＿＿＿＿＿＿＿＿

DOS, Windows, UNIX, Linux　　＿＿＿＿＿＿＿＿＿＿

Oracle, MySQL, SQL Server, DB2　＿＿＿＿＿＿＿＿＿＿

Microsoft Word, WPS Office, Notepad　＿＿＿＿＿＿＿＿＿＿

Photoshop, CorelDRAW　　　　　＿＿＿＿＿＿＿＿＿＿

What kind of software may we use? Match the software with the activities.

word processor　　　computer-aided design software　　　image processing software

translation tool　　　electronic reading tool　　　input method editor

Lesson Eleven

Read and say.

- What are the functions of archive manager?
  It is used to compress or decompress files and to back up data.
- How many kinds of archive managers are there?
  There are a lot, such as WinRAR, WinZIP, 7-ZIP, ARJ and JAR.
- Which are the most commonly used?
  WinRAR and WinZIP. They support different file formats.

WinRAR

WinZIP

Act out the dialogue.

Tom: I want to email many pictures to my friend, but the files are too large. What can I do?

Wang Zixuan: You may use the archive manager like WinRAR to compress files.

Tom: But how can I do it? Could you show me on my computer?

Wang Zixuan: Certainly. First select the file that you would like to compress. Right-click the mouse and choose "**Add to archive**..." command from the pop-up menu.

Tom: Do I need to name it?

Wang Zixuan: WinRAR suggests a default name. But you can change it and also add a path.

Tom: What should I do next?

Wang Zixuan: At last, press the "OK" button to start archiving. Look, it's easy.

Tom: Yes, thank you.

Wang Zixuan: You're welcome.

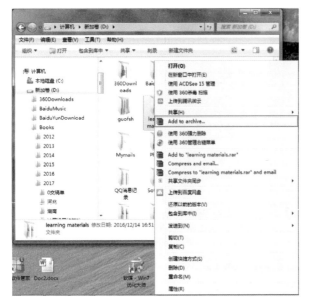 

Word Bank

扫一扫
听录音

archive /ˈɑːkaɪv/ n. 档案文件
              v. 存档
decompress /ˌdiːkəmˈpres/ v. 解压缩
format /ˈfɔːmæt/ n. 格式
suggest /səˈdʒest/ v. 建议
add /æd/ v. 添加
press /pres/ v. 按下
pop-up menu 快捷菜单
archive manager 档案文件管理器

manager /ˈmænɪdʒə(r)/ n. 管理器
compress /kəmˈpres/ v. 压缩
commonly /ˈkɒmənli/ adv. 通常地
choose /tʃuːz/ v. 选择
default /dɪˈfɔːlt/ n. 默认
path /pɑːθ/ n. 路径
button /ˈbʌt(ə)n/ n. 按钮
back up 备份
file format 文件格式

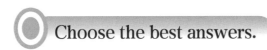 **Choose the best answers.**

1. What kind of archive manager are you using now?
    WinRAR   WinZIP   7-ZIP   ARJ   JAR

2. What are the functions of archive manager?
    ( ) compress files   ( ) design software
    ( ) decompress files   ( ) backup data

Put the following sentences in the right order.

(　　) → (　　) → (　　) → (　　) → (3) → (　　)

1. Press the "OK" button to start archiving.
2. Choose "Add to compressed files" from the **pop-up menu**.
3. Choose a compression method.
4. Select and right-click the files that you would like to compress.
5. Give a name and the path to the compressed file.
6. Choose the type of the archive format.

Lesson Twelve

Tell the full names.

CAD _____   DBMS _____

WPS _____   CAI _____

PC _____   CAM _____

Put the followings into the right categories.

| Windows | Photoshop | Office | Flash | Oracle |
| MySQL | CorelDRAW | Adobe Reader | UNIX | QQ |

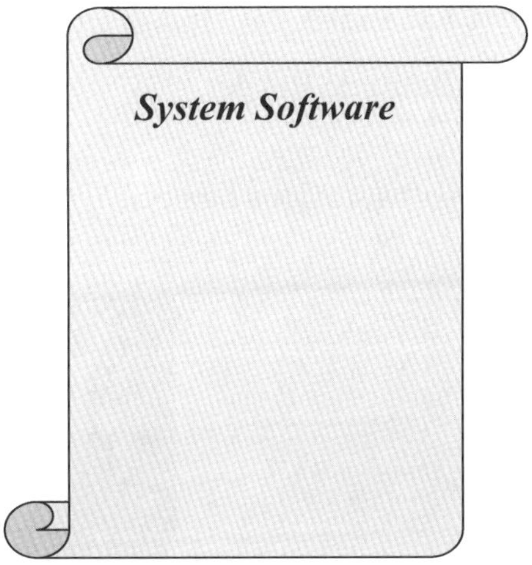

System Software

Application Software

◉ Translate the words in the table into Chinese with the help of the interface.

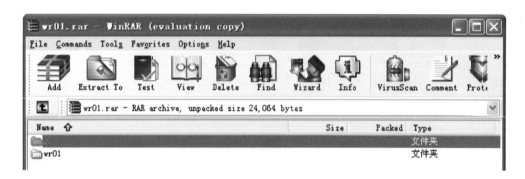

 Add	 Info	 Delete	 View
 Extract To	 Repair	 Find	 Lock

查看　信息　添加　　加锁
修复　查找　解压到　删除

◉ Please help the people choose the proper software.

| Audio Player | Video Displayer | Chat Tool | Image Editor |
| Total Recorder | Video Editor | Electronic Reader | |

1. Lin Tao wants to record his voice. What kind of software does he need?

   The proper software: _____.

2. The Mike family is on a video call with Lucy's family. What kind of software do they need?

   The proper software: _____.

3. Mary wants to make the photo clearer and nicer. What kind of software does she need?

   The proper software: _____.

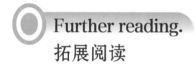

Further reading.
拓展阅读

Harmony OS is an open-source operating system designed for various devices and scenarios. It first launched on Internet-of-Things (IoT) devices, including wearable and tablets.

As a next-generation operating system for smart devices, Harmony OS provides a common language for different kinds of devices to connect and collaborate, providing users with a more convenient, smooth, and secure experience.

Harmony OS greatly enhances the interactive speed between devices and improves the efficiency of their computing power, thus providing customers with a more optimized cross-device user experience.

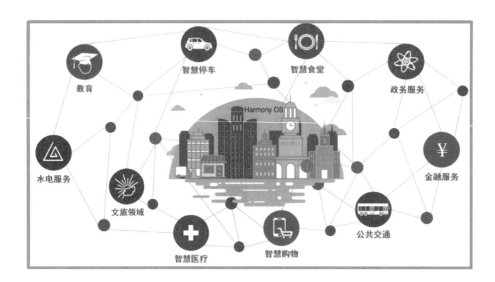

▼ **Discuss:** What kind of society Harmony OS brings?

Hints: harmonious 和谐，more secure 更安全，revolution 革命，confident 自信，historic 历史悠久，intelligent 充满智慧的，noble 崇高无私，ambitious 雄心壮志，national confidence 民族自信，national self-improvement 民族自强，national pride 民族自豪

Unit 4 Operating System

Lesson Thirteen

Match the pictures with the operating systems.

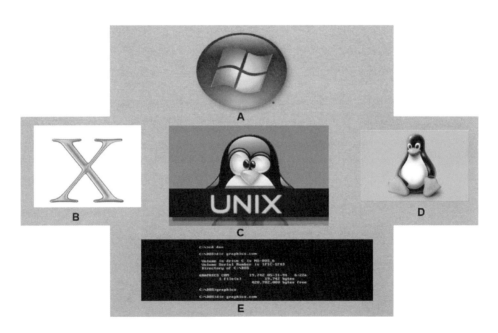

DOS _____ Windows _____ UNIX _____

Linux _____ Mac OS X _____

Match the operating systems with the right companies or people.

Microsoft Corporation	MS-DOS
Linux Torvalds	Windows
Apple Incorporated	Unix
AT&T Bell Laboratories	Linux
	Mac OS

 Listen and fill in the blanks.

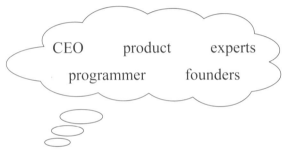

- In 2022, Xiaomi Incorporated succeeded in producing a new generation of _____ (产品) —— Xiaomi 13.
- Steve Jobs once was the _____ (首席执行官) of Apple Incorporated and Pixar Animation Studios.
- Bill Gates and Paul Allen are the _____ (创始人) of Microsoft, which is the world leader in the PC software development.
- Linus Torvalds is a famous computer _____ (程序员) and the inventor of Linux kernel.
- AT&T Labs is made up of 1,300 scientists, engineers and _____ (专家).

Word Bank

incorporated /ɪnˈkɔːpəreɪtɪd/ adj. (公司) 股份有限的
succeed /səkˈsiːd/ v. 成功
founder /ˈfaʊndə(r)/ n. 创始人
programmer /ˈprəʊɡræmə(r)/ n. 程序员

laboratory /ləˈbɒrətri/ n. 实验室
studio /ˈstjuːdɪəʊ/ n. 工作室
development /dɪˈveləpmənt/ n. 发展，开发
inventor /ɪnˈventə(r)/ n. 发明者

engineer /ˌendʒɪˈnɪə(r)/ n. 工程师　　expert /ˈekspɜːt/ n. 专家

be made up of　由……组成

Lesson Fourteen

 Discussion.

1. Which operating system is more widely used in China?
2. Do you know the development of Windows?

 Reading.

扫一扫
听录音

The Windows OS

One of the most widely used operating systems is Microsoft Windows. Windows is a popular operating system(OS) for personal computers today. It has many versions over the years.

Windows Family

Windows 2000
（In 2000）

Windows Me
（In 2000）

Windows XP
（In 2001）

Windows 2000 is a network operating system for business use and server computers.

Windows Me is the first operating system to support Universal Plug and Play(UPnP), using standard Internet protocols to connect PCs.

Windows XP can switch between user accounts without rebooting or closing open programs. Some new features are related to multimedia and communications.

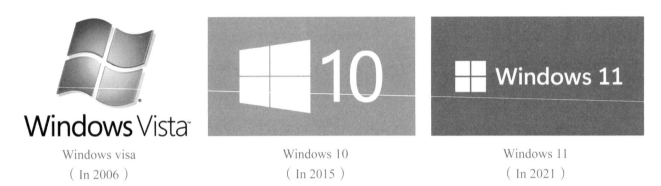

| Windows visa | Windows 10 | Windows 11 |
| (In 2006) | (In 2015) | (In 2021) |

Windows Vista has more functions, especially graphic processing function, Windows Aero and Windows Indexing Service.

Windows 7 is still the most popular new operating system, and it is used to rescue the failed Windows Vista.

Windows 10 is a revolutionary change of the operating system. It makes computer operation more convenient, and offers an easy working environment for people.

Windows 11 is applied to devices such as computers and tablets, focusing on improving the work efficiency of end users in a flexible and ever-changing experience.

Word Bank

扫一扫
听录音

widely /'waɪdli/ adv. 广泛地
graphical /'græfɪkl/ adj. 图形的
degree /dɪ'griː/ n. 程度
server /'sɜːvə(r)/ n. 服务器
protocol /'prəʊtəkɒl/ n. 协议
switch /swɪtʃ/ v. 转换
feature /'fiːtʃə(r)/ n. 特征，特性
communication /kə,mjuːnɪ'keɪʃ(ə)n/ n. 交流，通信
rescue /'reskjuː/ v. 挽救
revolutionary /,revə'luːʃənəri/ adj. 革命性的

version /'vɜːʃ(ə)n/ n. 版本
interface /'ɪntəfeɪs/ n. 界面，接口
integration /,ɪntɪ'greɪʃ(ə)n/ n. 集成；一体化
standard /'stændəd/ adj. 标准的
connect /kə'nekt/ v. 连接
reboot /,riː'buːt/ v. 重新启动（计算机系统）
related /rɪ'leɪtɪd/ adj. 相关的
failed /feɪld/ adj. 失败的
convenient /kən'viːnjənt/ adj. 方便的

environment /ɪnˈvaɪrənmənt/ n. 环境
be related to 与……有关

Universal Plug and Play(UPnP) 通用即插即用

 Unscramble the words and find the correct meanings.

dda	→	add
unnoitcf	→	_____
erinovs	→	_____
oncenvteni	→	_____
eaeurtf	→	_____
eesucr	→	_____

方便的
增加
版本
挽救
功能
特色

 Match the different Windows versions with the relative information.

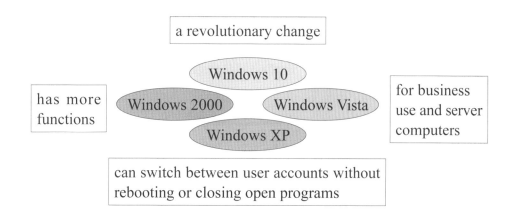

Lesson Fifteen

Read and say.

- What operating systems do mobile phones commonly use?
  They are iOS, Android and Harmony OS.
- What's the advantage of Android operating system?
  It's more widely used and it is free to download software.
- What's the advantage of Harmony OS?
  It is stable and fast.

Act out the dialogue.

Li Lei: Linda, I'm so sad that my mobile phone doesn't work.

Linda: What a pity. But it's a good chance for you to buy a new one.

Li Lei: Yes, but which type of mobile phone is better?

Linda: Huawei Mate 50 Pro is better.

Li Lei: What kind of operating system does Mate 50 Pro use?

Linda: It's Harmony OS. The system is very stable and safe.

Li Lei: How about the Apple?

Linda: The signal is weak in many places.

Li Lei: I will choose Mate 50 Pro. Any other advantages about the Harmony OS?

Linda: The Harmony OS is widely used in smartphone platform.

Li Lei: Why is it widely used?

Linda: Maybe because Harmony OS provids a common language for different kinds of devices to connect.

Li Lei: I see. Thank you. Would you like to go with me if I decide to buy it?

Linda: Yes, I'd love to.

Word Bank

advantage /əd'vɑːntɪdʒ/ n. 优点
free /friː/ adj. 免费的
stable /'steɪb(ə)l/ adj. 稳定的
suitable /'suːtəb(ə)l/ adj. 合适的

platform /'plætfɔːm/ n. 平台
download /ˌdaʊnləʊd/ v. 下载
safe /seɪf/ adj. 安全的
signal /sɪgnəl/ n. 信号

 Choose the suitable operating system for the following people.

1. I am seventeen years old. I like to download a lot of software into my mobile phone for free. What kind of operating system should I choose?
_____
_____

2. I am a merchandiser of a foreign trade corporation. I need to receive and send emails on business to different cities. My phone signal is not good. Which operating system should I choose?
_____
_____

3. I am an office lady, and I like listening to music and reading novels after work. What kind of operating system should I choose?
_____
_____

Lesson Sixteen

Match the abbreviations in Column A with their full names in Column B.

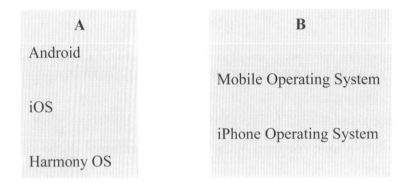

Fill in the blanks in the Windows desktop with the given names.

- **Desktop icon** — opening a program, document, or folder.
- **Start button** — opening the Start menu.
- **Start menu** — starting the programs.
- **Taskbar** — at the bottom of the desktop.
- **A window** — a box containing the program, document, or other data.

Fill in the blanks with the proper words.

> toolbar　Restart　Double-click　scroll　select
> Click　Insert　Start menu　installation　agreement

(1) *How to install Windows?*

Instructions:

1. _____（插入）the Windows installation DVD into your DVD - ROM drive.
2. _____（重启）your computer.
3. Press any key when you get the prompt "Press any key to boot from DVD."
4. Press the Enter key when you get to the "Welcome to Setup" screen.
5. Read the _____（协议）. Press the Page Down key to _____（滚动）down to the bottom.
6. Press the Enter key to _____（选择）"Unpartitioned space"（未区分空间）.
7. Press the Enter key again.
8. Click on the "Customize" button and select the preferred language in the drop-down menu.
9. Follow the rest of the prompts to finish the _____（安装）.
10. Restart your computer when the installation finished.

(2) *How to open a window or start a program?*

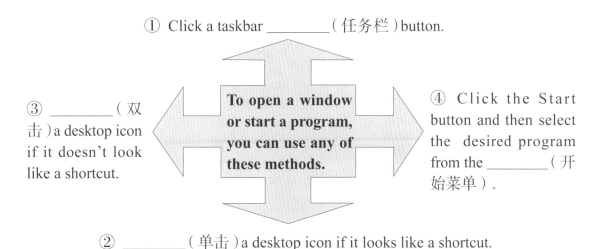

① Click a taskbar _____（任务栏）button.

③ _____（双击）a desktop icon if it doesn't look like a shortcut.

To open a window or start a program, you can use any of these methods.

④ Click the Start button and then select the desired program from the _____（开始菜单）.

② _____（单击）a desktop icon if it looks like a shortcut.

Put the following steps in the right order.

Learn to create a new folder. You can follow these steps:

(　　) → (　　) → (　　)

1. Name the folder, simply type in the name and then press the enter key.

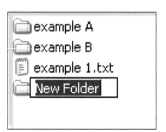

2. Right click anywhere in the white space（空白处）.

3. Move the mouse over "New" in the pop-up menu. Click on "Folder" from the sub-menu as shown in the following picture.

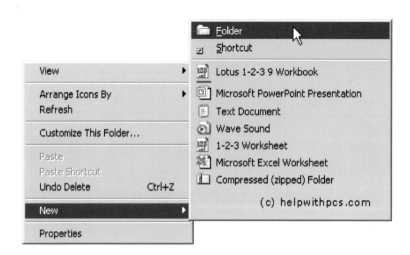

Further reading.
拓展阅读

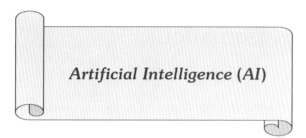

Artificial Intelligence (AI)

AI has played a remarkable role in helping a slew of industries fight the negative effects of the pandemic, especially in manufacturing, transportation, logistics, medical care and education.

The country has been ramping up its AI development efforts. In its 14th Five-Year Plan (2021—2025) proposals, the nation has highlighted the role of frontier technologies like AI, 5G, super computing and quantum computing in development.

In the world's top artificial intelligence challenge OGB (Open Graph Benchmark), 360 AI Research Institute, relying on the "domestic self-developed" knowledge expression model, topped the list of difficult tasks OGB-wikikg2. It is reported that this is the first time that Chinese digital security enterprises have topped the list.

▼ **What fields can artificial intelligence be applied to?**

☐ self-driving（自动驾驶）　　☐ self-service bank（自助银行）
☐ mechanical arm（机械臂）　　☐ health care（医疗保健）
☐ educational field（教育领域）

▼ **Discuss:**

Do you think robots will do all the work instead of people in the future?

Unit 5 Office Automation

Lesson Seventeen

◉ Do you know the following programs?

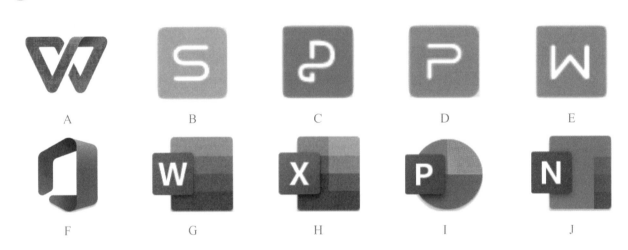

◉ Match the names with the pictures above.

Microsoft Office _____   WPS Office _____

Microsoft Word _____   WPS Presentation _____

Microsoft PowerPoint _____   WPS Spreadsheet _____

Microsoft Excel _____   WPS PDF _____

Microsoft OneNote _____   WPS Writer _____

Match the following programs with their names and functions.

Microsoft Office	Functions	WPS Office
( ) 1. Microsoft Word	A. process spreadsheets	( ) 5. WPS Writer
( ) 2. Microsoft Excel	B. process words	( ) 6. WPS Presentation
( ) 3. Microsoft PowerPoint	C. present graphic documents	( ) 7. WPS Spreadsheet
( ) 4. Microsoft OneNote	D. gather users' notes	( ) 8. WPS PDF
	F. read and edit on PDF files	

Listen and fill in the blanks.

扫一扫 听录音

features   data   slides
documents   notebook

As popular office APPs, WPS Writer and Microsoft Word both can allow us to create _____（文档）in various formats and provide complete features to meet most of our office needs.

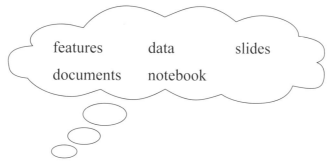

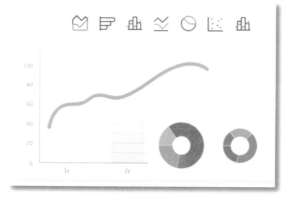

WPS Spreadsheet provides massive free charts to help us visualize our _____（数据）with ease.

WPS Presentation provides various delicate templates with different themes. We can download them for free to beautify our _____（幻灯片）.

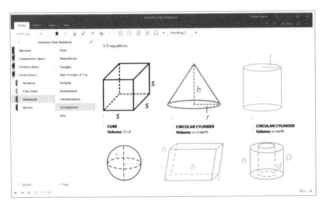

Microsoft OneNote is a cross-functional digital _____（笔记本）for all our note-taking needs, so it is easy for us to take notes during the class.

WPS PDF provides _____（功能）such as image and text editing, page cropping, and header and footer settings. We can edit PDF files in the same way as we would do in a Word document processor.

Word Bank

扫一扫
听录音

presentation /ˌprez(ə)nˈteɪʃ(ə)n/ n. 演示文稿
process /ˈprəʊses/ v. 处理
document /ˈdɒkjumənt/ n. 文档
edit /ˈedɪt/ v. 编辑
template /ˈtempleɪt/ n. 样板
crop /krɒp/ v. 剪短；剪裁

spreadsheet /ˈspredʃiːt/ n. 电子表格
graphic /ˈɡræfɪk/ adj. 图解的
gather /ˈɡæðə(r)/ v. 收集
chart /tʃɑːt/ n. 图表
digital /ˈdɪdʒɪt(ə)l/ adj. 数字的
processor /ˈprəʊsesə(r)/ n. 处理器

Lesson Eighteen

1. How many kinds of software for office automation do you know?
2. Which is more commonly used in your life? Why?

Office automation refers to the varied computer software, which is used to deal with office information.

Whether people are in the office, school, or at home, WPS Office can always meet their requirements for instant creativity and collaboration in daily-life scenarios. It will always help people convey passion, inspire innovation, and stand out among peers, whether they are a business professional or a student.

WPS Office is available on multiple platforms. People can work anytime and anywhere on their mobile phone or computer. WPS Office is an office suite for Microsoft Windows, Mac, Linux, iOS, Android and Harmony, developed by Chinese software developer Kingsoft.

WPS means word processing system. WPS Office is a lightweight, feature-rich comprehensive office suite with high compatibility. WPS Office allows people to edit

45

files in Writer, Presentation, Spreadsheet and PDF to improve their work efficiency.

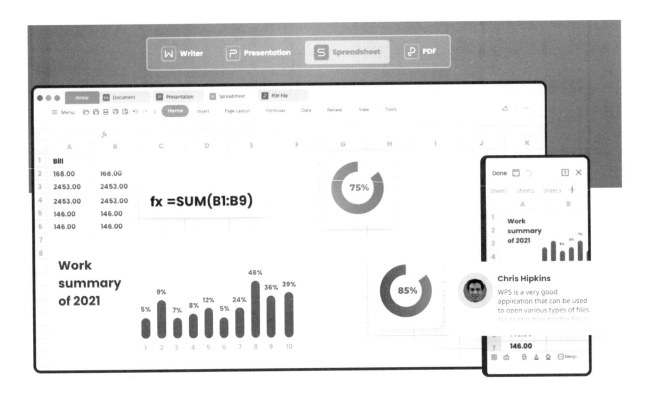

WPS Office Software is a leading office productivity suite for PC and mobile. It is a high-performance and cost-effective office suite fully compatible with Microsoft PowerPoint, Excel, and Word. As a leading Internet service and software company, it independently develops main products and services and owns over 941 independent intellectual property rights.

Word Bank

requirement /rɪˈkwaɪəmənt/ n. 要求
inspire /ɪnˈspaɪə(r)/ v. 赋予灵感
peer /pɪə(r)/ n. 同龄人
platform /ˈplætfɔːm/ n. 平台
feature-rich /ˈfiːtʃ ə(r) rɪtʃ/ adj. 丰富的
suite /swiːt/ n. 套装软件
improve /ɪmˈpruːv/ v. 改进

convey /kənˈveɪ/ v. 传送
innovation /ˌɪnəˈveɪʃ(ə)n/ n. 创新
professional /prəˈfeʃn(ə)l/ n. 专业人士
lightweight /ˈlaɪtweɪt/ adj. 轻量的
comprehensive /ˌkɒmprɪˈhensɪv/ adj. 综合的
compatibility /kəmˌpætəˈbɪləti/ n. 兼容性
efficiency /ɪˈfɪʃ(ə)nsi/ n. 效率

Choose the best answer according to the text.

1. Office automation refers to the varied _____, which is used to deal with office information.
   A. computer memory   B. computer software   C. computer hardware
2. WPS Office was designed by _____.
   A. Microsoft   B. Kingsoft   C. Star
3. WPS Office allows people to edit files in Writer, _____, Spreadsheet and PDF to improve their work efficiency.
   A. Word   B. Presentation   C. PowerPoint
4. WPS Office is compatible with Microsoft PowerPoint, _____, and Word.
   A. Writer   B. Excel   C. Spreadsheet

Fill in the banks with the correct words.

| information | requirements | platforms | edit | suite |

1. Office automation refers to the varied computer software, which is used to deal with office _____.
2. Whether people are in the office, school, or at home, WPS Office can always meet their _____ for instant creativity and collaboration in daily-life scenarios.
3. WPS Office is available on multiple _____.
4. WPS Office allows people to _____ files in Writer, Presentation, Spreadsheet and PDF to improve their work efficiency.
5. WPS Office is an office _____ for Microsoft Windows, Mac, Linux, iOS, Android and Harmony, developed by Chinese software developer Kingsoft.

Lesson Nineteen

Read and match.

Word processing means "to create, edit, save, and print written documents". Do you know how to use it?

- Creating and editing documents
- Scrolling and moving the Insertion Point
- Line breaks and paragraph breaks
- Editing text
- Spell-checking and other editing tools
- Documents formatting
- Characters formatting
- Paragraphs formatting

- 编辑文本
- 创建和编辑文档
- 滚屏和移动插入点
- 字符格式化
- 段落格式化
- 拼写检查和其他编辑工具
- 文档格式化
- 分行和分段

Act out the dialogue.

Linda: Can you do me a favor?

Tom: Sure, what is it?

Linda: I'd like to know something about files. First of all, what is a file?

Tom: A file is a collection of data that is stored together.

Linda: What can we do with files?

Tom: You can create, name, rename, save or delete files.

Linda: How many file types do you know?

Tom: Computer files come in different types, such as text files, graphic files, program files and email files.

Linda: Why is a period used in the file name?

Tom: A period is used to separate the name of the file from the type or extension. For example, a Word file has the extension name ".docx".

Linda: Where can I find a file if I delete it carelessly?

Tom: You can recover a deleted file in the Recycle Bin.

Linda: Now I have got some ideas about files. Thanks for your help.

Tom: You're welcome.

Word Bank

create /krɪˈeɪt/ *v.* 创建

print /prɪnt/ *v.* 打印

delete /dɪˈliːt/ *v.* 删除

separate /ˈseprət/ *v.* 分隔

pathway /ˈpɑːθweɪ/ *n.* 路径

carelessly /ˈkeələsli/ *adv.* 粗心地

save /seɪv/ *v.* 节省；保存

collection /kəˈlekʃ(ə)n/ *n.* 集合

period /ˈpɪəriəd/ *n.* 句点

extension /ɪkˈstenʃ(ə)n/ *n.* 扩展

folder /ˈfəʊldə(r)/ *n.* 文件夹

recover /rɪˈkʌvə(r)/ *v.* 恢复

 Translate the menu names of WPS Writer into English.

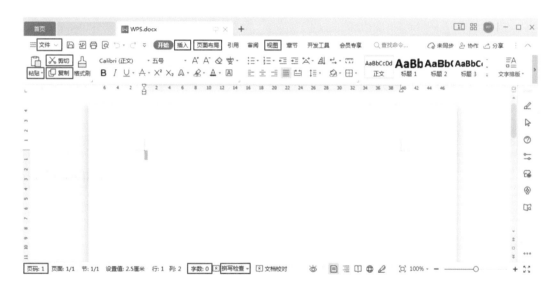

Insert	File	Paste	Page Layout	View
Cut	Copy	Page Num	words	Spell Check
文件 _____	插入 _____	页面布局 _____	视图 _____	粘贴 _____
剪切 _____	复制 _____	页码 _____	字数 _____	拼写检查 _____

Learn to use WPS Writer and fill in the blanks.

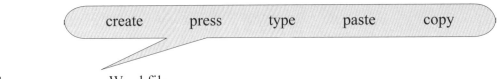

create   press   type   paste   copy

1. _____ a Word file.

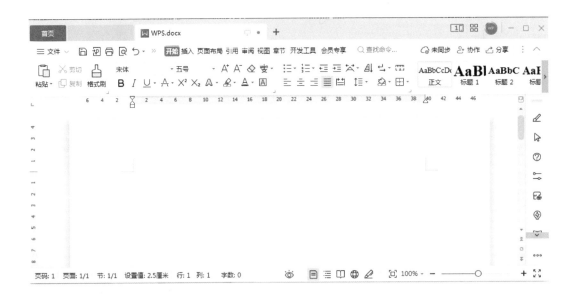

2. _____ the text.

3. Put the cursor（光标）at the beginning of the words you want, _____ the left button of the mouse, drag to the end of the words.

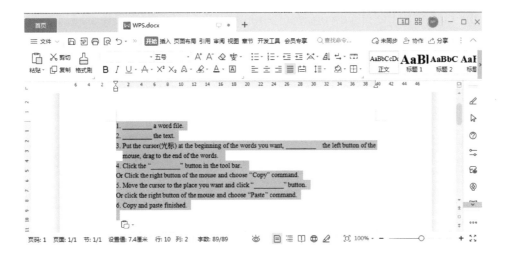

4. Click the "_____" button in the tool bar.

Or click the right button of the mouse and choose "Copy" command.

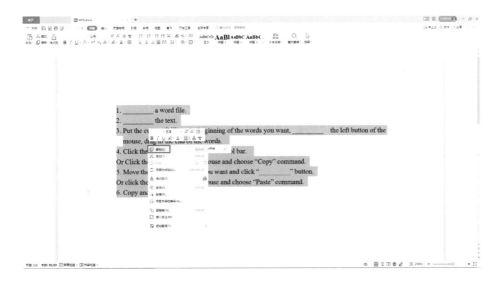

5. Move the cursor to the place you want and click "_____" button.

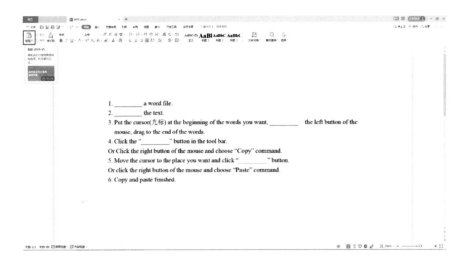

Or click the right button of the mouse and choose "Paste" command.

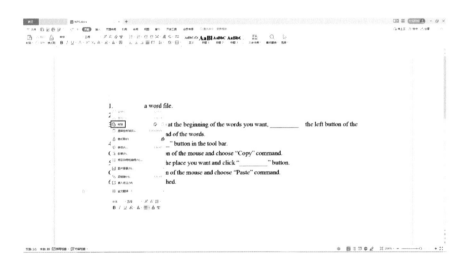

6. Copy and paste finished.

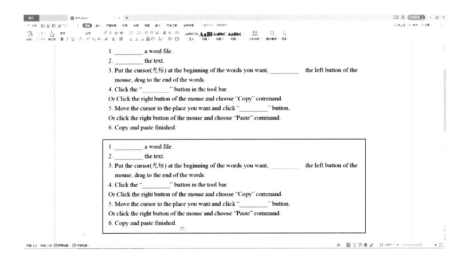

Lesson Twenty

 Fill in the blanks with the proper words.

How to create a Word document?

1. _____ a blank document.
2. _____ the text.
3. _____ the document.
4. _____ the document.

How to create a presentation?

1. _____ a blank presentation.
2. _____ a design template.
3. _____ new slides.
4. _____ text and insert clip art.
5. _____ the presentation.
6. _____ the presentation.

How to create a spreadsheet?

1. _____ a blank worksheet.
2. _____ the cell.
3. _____ numbers.
4. _____ data.

5. _____ formulas to calculate.
6. _____ the worksheet.

Tell the functions of the following buttons of WPS Office in Chinese.

Functions of Buttons

▢	create a new blank document	📂	open a file
💾	save the active file	🖨	print the active file
🔍	print preview	abc✓	spelling and grammar check
✂	cut	🗐	copy
📋	paste	🖌	format painter
↶	undo or reverse the last command	↷	redo or reverse the action of the Undo button

Carefully review the functions of the buttons in WPS Spreadsheet and take the short quiz below.

Which button do you select?

	A	B	C	D	
1	🗐	abc✓	📂	📋	Check spelling and grammar.
2	Σ	✂	A↓	🔍	Remove the selection from the active document.
3	📊	🖌	🖨	💾	Changes are to be saved.
4	abc✓	Σ	A↓	📊	Get a quick sum of numbers.
5	🖨	⊙	↶	A↓	Print the worksheet.
6	💾	▢	abc✓	📂	Open a worksheet.
7	A↓	🔍	📊	✂	See the worksheet before printing it.

54

Learn how to use WPS Presentation.

> templates　　blank　　desktop　　layout　　Slide

1. Starting a file. Double click on the WPS Presentation on the _____.
2. Create a _____（空白）presentation.

3. After you click the "Design" button in the tool bar, you will see many free design _____ and you can choose what you like.

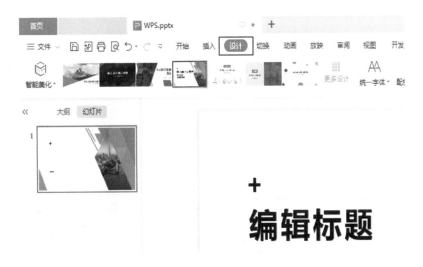

4. After click the "Layout" button, a window pops up（弹出）, asking you to select the _____ of the first slide.

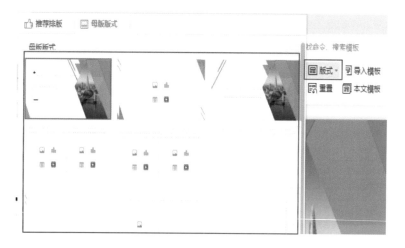

5. Different views that WPS Presentation demonstrates（演示）.

Normal View   Slide Sorter View   _____ Show View

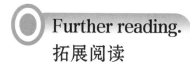

Further reading.
拓展阅读

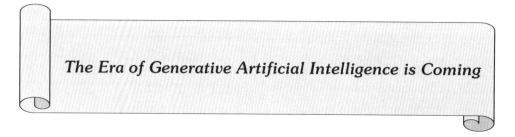

Generative AI technology not only generates high-quality content such as texts, images and music, but also can be used in automatic programming, machine learning, data generation, game development, virtual reality and other fields. In China, Baidu has launched "ERNIE Bot"（文心一言）, Alibaba has released "Tongyi Qianwen"（通义千问）, while Tencent, 360, Huawei and other Internet giants have announced their layout in generative AI services. It has already started changing our way of life and work, creating new opportunities for the development of different industries.

▼ **Which fields can Generative AI technology be applied to?**

☐ generates high-quality and diverse content
☐ automatic programming
☐ machine learning
☐ data generation
☐ game development
☐ virtual reality

▼ **Discuss:**

What can AI do for us?

Unit 6 The Internet

Lesson Twenty-one

Dick the things you do with the Internet.

- ☐ chat with friends
- ☐ download music and films
- ☐ watch films
- ☐ play computer games
- ☐ search for information
- ☐ read novels
- ☐ send and receive emails
- ☐ learn English
- ☐ listen to music
- ☐ buy things
- ☐ learn how to use softwares
- ☐ read news

Give examples based on different types of websites.

Different types of websites	Examples
mass media website	
search engine	
IT website	
entertainment website	
educational website	
e-commerce website	
online shopping website	

Listen and fill in the blanks.

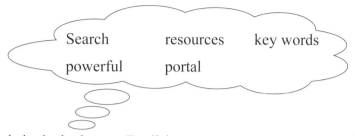

- China Daily website is the largest English _____ (门户网站) in China, providing news, business information, BBS and learning materials.
- _____ (搜索) through the website to get website services.
- Many learning _____ (资源) can be downloaded from educational websites.
- After we get into the search engine, input the _____ (关键词) to be searched, press "enter" and then the automatic search will begin.
- Baidu is considered as the most _____ (强大的) search engine in China.

Word Bank

search /sɜːtʃ/ v. 搜索
entertainment /ˌentəˈteɪnmənt/ n. 娱乐
commerce /ˈkɒmɜːs/ n. 贸易，商业
provide /prəˈvaɪd/ v. 提供
international /ˌɪntəˈnæʃ(ə)nəl/ adj. 国际的
consider /kənˈsɪdə(r)/ v. 认为
mass media 大众传媒

engine /ˈendʒɪn/ n. 引擎，发动机
education /ˌedʒuˈkeɪʃ(ə)n/ n. 教育
portal /ˈpɔːtl/ n. 门户网站
exceptional /ɪkˈsepʃən(ə)l/ adj. 杰出的
resource /rɪˈsɔːs/ n. 资源
powerful /ˈpaʊəf(ə)l/ adj. 强大的

Lesson Twenty-two

Discussion.

1. How often do you surf the Internet?
2. What can you do on the Internet?

Reading.

Internet for the Better Life

What is the Internet? It is a global computer network. What can we do on the Internet in our daily life? We can send and receive emails, or go job hunting. We can also buy goods from online shopping websites. What's more, we can pay our fees online within doors. Besides, the Internet provides us with information and entertainment. It links together millions of computers around the world! It becomes more and more powerful. With the development of computer science, China is booming its digital economy.

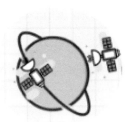

Since BeiDou-3 navigation satellite system starts providing global service in July 2020, there are over 1 billion terminal products with BeiDou positioning functions, including over 7.8 million automobiles and 100,000 automatic driving farming vehicles.

Smart industry, smart transportation, intelligent health, smart energy and other fields have become areas where the number of industrial Internet of Things connections is growing rapidly.

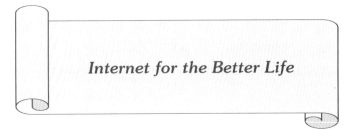

Areas of rapid growth in the number of industrial IoT connections:

Smart industry   Smart transportation

Intelligent health   Smart energy

The Internet also contributes to the digital transformation of agriculture.

It also flourishes e-commerce, including online retail sales of comsumer goods, trade in digitally deliverable services and turnover of cross-border e-commerce.

Smart Education of China

It does benefit to IT application in education. According to the statistics of 2021, all primary and middle schools have Internet access.

Over 210, 000 schools have wireless network services and 99.5% of all schools have multimedia classrooms.

With the Internet, our life is becoming more and more wonderful.

Word Bank

global /'gləʊb(ə)l/ *adj.* 全球的
million /'mɪljən/ *n.* 百万
economy /ɪ'kɒnəmi/ *n.* 经济
satellite /'sætəlaɪt/ *n.* 人造卫星
terminal /'tɜːmɪn(ə)l/ *n.* 终端
vehicle /'viːəkl/ *n.* 交通工具
transportation /ˌtrænspɔː'teɪʃ(ə)n/ *n.* 运输
transformation /ˌtrænsfə'meɪʃ(ə)n/ *n.* 转型
flourish /'flʌrɪʃ/ *v.* 繁荣
turnover /'tɜːnəʊvə(r)/ *n.* 年营业额
statistic /stə'tɪstɪk/ *n.* 统计数字
BeiDou positioning function 北斗定位功能
contribute to 有助于

fee /fiː/ *n.* 费用
boom /buːm/ *v.* 迅速发展
navigation /ˌnævɪ'geɪʃ(ə)n/ *n.* 导航
billion /'bɪljən/ *n.* 十亿
automatic /ˌɔːtə'mætɪk/ *adj.* 自动的
industry /'ɪndəstri/ *n.* 工业
intelligent /ɪn'telɪdʒənt/ *adj.* 智能的
agriculture /'ægrɪkʌltʃə(r)/ *n.* 农业
retail /'riːteɪl/ *n.* 零售
application /ˌæplɪ'keɪʃ(ə)n/ *n.* 应用
multimedia /ˌmʌlti'miːdiə/ *n.* 多媒体
Internet of Things 物联网

 Choose the best answer according to the text.

1. What is the Internet?
   A. It's a worldwide web station.
   B. It's a worldwide network.
   C. It's an information server.

2. BeiDou-3 global navigation satellite system starts providing global service in_____.
   A. July 2020   B. July 2019   C. June 2020

3. Smart industry, smart transportation, _____, smart energy and other fields have become areas where the number of industrial Internet of Things connections is growing rapidly.
   A. intelligent health   B. smart homes   C. self-driving

4. What's the meaning of "IT application in education"?
   A. Having internet access.
   B. Having wireless network services.
   C. Having traditional classrooms.

Complete the sentences with the given words.

smart	digital	positioning	retail	multi-media	access

1. China is booming the _____ economy.
2. There are over 1 billion terminal products with BeiDou _____ functions.
3. China Mobile has established an institute to build 5G network, develop _____ transportation and industrial application.
4. Last month, _____ sales of consumer goods climbed 3.3 percent year on year, up from a 0.5 percent growth in August.
5. 99.5% of all schools have _____ classrooms.
6. Any country who gained full _____ to the Internet should be committed to promoting Internet development and governance.

Discuss and complete the chart according to the text.

Do you like surfing the Internet? What are the advantages and disadvantages of the Internet?

Advantages	Disadvantages
1. We can contact friends by e-mail.	1. Students may spend too much time on online games.
2.	2.
3.	3.
4.	4.
…	…

What is your suggestion to use the Internet?

Lesson Twenty-three

Read and say.

Digital transformation of agriculture

IoT

BIG DATA

5G

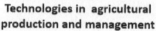
Technologies in agricultural production and management

AI

Act out the dialogue.

Linda: Li Lei, what do you think of farming?

Li Lei: Farming is very important in my hometown. People have a huge demand for agricultural products.

Linda: What are the main agricultural products in your hometown?

Li Lei: Main crops are seasonal fruits, such as litchi, banana, mango and so on. Rice is produced in my hometown, too.

Linda: Wow, there are so many products. It must be managed by many people on farms.

Li Lei: Thanks to the development of Internet, agriculture has become quite modernized and farm machines have taken the place of farm

Water-holding capacity Optimum moisture Wilting point

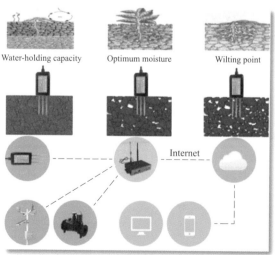

Internet

workers. We call it smart agriculture.

Linda: Could you tell me more about it?

Li Lei: Of course. It is the combination of Internet of Things technology and agriculture. For example, smart irrigation system is to monitor the environment and realize smart irrigation with the help of Internet.

Linda: It has changed the way of agricultural production, hasn't it?

Li Lei: Yes, it has.

扫一扫
听录音

Word Bank

huge /hjuːdʒ/ *adj.* 巨大的
agricultural /ˌæɡrɪˈkʌltʃərəl/ *adj.* 农业的
modernize /ˈmɒdənaɪz/ *v.* 使现代化
irrigation /ˌɪrɪˈɡeɪʃ(ə)n/ *n.* 灌溉
realize /ˈriːəlaɪz/ *v.* 实现
take the place of 代替，取代
smart irrigation system 智能灌溉系统

demand /dɪˈmɑːnd/ *n.* 需要
seasonal /ˈsiːzən(ə)l/ *adj.* 节令性的
combination /ˌkɒmbɪˈneɪʃ(ə)n/ *n.* 结合体
monitor /ˈmɒnɪtə(r)/ *v.* 监视
thanks to 幸亏；归因于
Internet of Things technology 物联网技术

 Unscramble the words and find the correct meanings.

mandde (demand)		A. 需要
meoderniz ( )		B. 巨大的
irigatiorn ( )		C. 产生
negerate ( )		D. 灌溉
hgue ( )		E. 使现代化
onitorm ( )		F. 监视

Choose the Internet application according to the different situation.

| A. self-driving | B. mobile payment | C. online classes |
| D. smart homes | E. long-distance operation | |

(   ) 1. During the epidemic, live broadcast software are used to have _____.

(   ) 2. The government believes that _____ technology will make the roads safer in the future.

(   ) 3. The _____ has become so popular that you just need to scan the QR code to finish your payment.

(   ) 4. Doctors can perform _____ through 5G technology.

(   ) 5. Some _____ can receive a visitor, allowing him to come in and offering him a drink.

Lesson Twenty-four

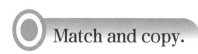 Match and copy.

Abbreviations	Full names	Chinese meanings
ISP	1. Internet Service Provider	A. 物联网
IE	2. Internet of Things	B. 因特网服务提供商
IoT	3. Internet Explorer	C. 因特网浏览器

ISP: _____ (   )

IoT: _____ (   )

IE: _____ (   )

Fill in the blanks.

| World Wide Web   search engine   IP address |

1. _____ is computer network consisting of a collection of Internet sites.

2. If Tom wants to find out a piece of news on electronic *China Daily*, he could use _____ to do it.

3. If the policemen want to find out the position of a computer, they should know the _____ of the computer.

Translate the steps to use on-line meeting software.

A. 登录在线会议应用软件；

B. 界面中有三个功能，分别是"加入会议""快速会议"和"预定会议"；

C. 下载并安装在线会议应用软件；

D. 单击"预定会议"按钮，设置会议主题、开始与结束时间等信息；

E. 单击"加入会议"按钮，输入会议号、你的姓名；

F. 单击"离开会议"按钮，可以暂时离开这个会议；

G. 单击"快速会议"按钮，可以创建一个会议。根据需要设置静音、开启视频、共享屏幕和管理成员等功能；

H. 单击"结束会议"按钮，可以结束这个会议。

(   ) 1. Download and install on-line meeting software;

(   ) 2. Sign up with your phone number or e-mail;

(   ) 3. There are three functions, namely "Join Meeting" "Quick Meeting" and "Book Meeting";

(   ) 4. Click "Book Meeting", set the information of the meeting theme, meeting date, start time and duration;

(   ) 5. Click "Join Meeting", type the meeting number and your name to join the meeting;

(   ) 6. Click "Quick Meeting" to create a meeting qucikly. Functions at the bottom, such as "Mute" "Open video" "Share screen" and "Manage members", can be set as required;

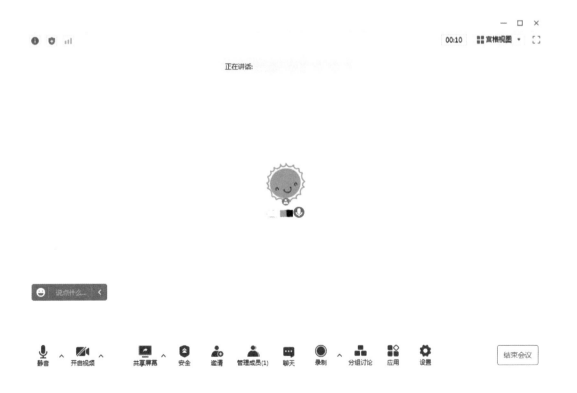

67

(   ) 7. Click "Leave Meeting" to leave the meeting temporarily；

(   ) 8. Click "End Meeting" to end the meeting.

Do online shopping in shopping website and fill in the blanks with proper words.

| account | payment | purchase | receiving | quality | refund |

Things to be prepared:

1. Register for an _____ （账户）on the online shopping website.

2. Register for a secured payment account.

Shopping online:

1. Choose goods after comparing price, freight（运费）, business reputation（店铺信誉） and other buyers' evaluations（评价）.

2. You can communicate with the seller for details.

3. Confirm the _____ （购买）. You can click on the "Immediate Purchase" and fill in the consignee（收件人）and receiving address information.

4. Finish the _____ （付款）according to the prompt（提示）. The payment for goods is put into the secured account.

5. Normally, after you receive the goods and click the "Confirmed _____ （收货）", the transaction is done.

6. There are some abnormal situations（异常情况）. If you can't receive the goods or the goods have _____ （质量）problems, you need to contact the seller or "Apply for _____ （退款）".

Further reading.
拓展阅读

Smart technology takes center stage at World Internet Conference

The Light of Internet Expo, part of the WIC, has attracted over 400 domestic and foreign companies both online and offline, with their latest innovations and cutting-edge technological applications on display. The concept of metaverse, 6G, human-computer interaction（人机交互）, internet security（互联网安全）, and industrial robot（工业机器人）is in the spotlight. Ele.me, one of China's food delivery platforms, is showcasing its new smart helmet for delivery workers to better protect their safety. China Mobile has released the metaverse digital human powered by optical motion capture technology. When a person is exercising or dancing, the digital figure can imitate the gesture and facial expression simultaneously.

▼ Which advanced technology application did you see during the 2022 World Internet Conference?

☐ metaverse ☐ 6G ☐ human-computer interaction
☐ internet security ☐ industrial robot

▼ How does the smart technology change our way of life and work?

Hints: Artificial intelligence（人工智能）, AR（增强现实）, VR（虚拟现实）, meta universe（元宇宙）, digital twins（数字孪生）

Unit 7　Multimedia

Lesson Twenty-five

Match the English names with the Chinese names according to the pictures.

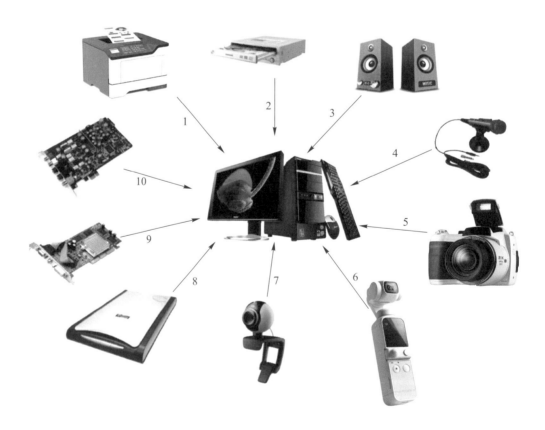

(　　) 1. 打印机　　　　　　A. digital camera

(　　) 2. 刻录机　　　　　　B. digital video camcorder

( ) 3. 音箱　　　　　　　　C. microphone

( ) 4. 麦克风　　　　　　　D. sound card

( ) 5. 数码相机　　　　　　E. video card

( ) 6. 数码摄像机　　　　　F. loudspeaker

( ) 7. 摄像头　　　　　　　G. CD-writer

( ) 8. 扫描仪　　　　　　　H. webcam

( ) 9. 视频卡　　　　　　　I. printer

( ) 10. 声卡　　　　　　　　J. scanner

Divide the devices into different categories.

digital camera	video camera	sound card	microphone
MIDI device	telephony system	loudspeakers	projector
CD-writer	monitor	printer	

Multimedia Devices	
Input Devices	
Output Devices	

Listen and fill in the blanks.

扫一扫
听录音

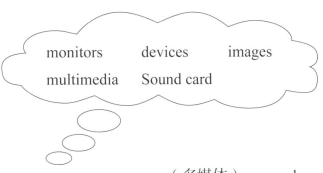

monitors　　devices　　images
multimedia　　Sound card

- MPC stands for _____（多媒体）personal computer.
- MPC has both input and output _____（设备）.
- _____（声卡）is an add-on device that generates analog sound signals from digital data.

- Digital cameras record digital _____(影像)on some types of digital storage medium.
- There are a lot of output devices for MPC, such as speakers, projectors, CD-writers, _____(显示器)and printers.

Word Bank

camcorder /ˈkæmkɔːdə(r)/ n. 便携式摄像机
sound /saʊnd/ n. 声音
telephony /təˈlefəni/ n. 电话（学）
generate /ˈdʒenəreɪt/ v. 生成
medium /ˈmiːdiəm/ n. 媒体

microphone /ˈmaɪkrəfəʊn/ n. 麦克风
loudspeaker /ˌlaʊdˈspiːkə(r)/ n. 音箱
projector /prəˈdʒektə(r)/ n. 放映机，投影仪
signal /ˈsɪgnəl/ n. 信号
stand for 代表

Lesson Twenty-six

1. How many types of multimedia information are there in our daily life?
2. What can we do with the help of multimedia?

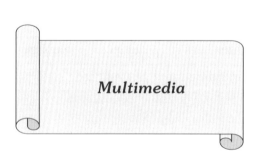

Multimedia plays an important role in computer development.

◆ The types of multimedia information.

Multimedia is the combination of text, sound, image, animation and video.

- Picture formats.

    GIF: Graphic Interchange Format.

    It can only store 256 or fewer colors.

    JPEG(or JPG): Joint Photographic Experts Group.

    It can store 16 million colors.

- Sound file formats.

    There are several types such as CD, MP3, WMA.

    MP3: MPEG Audio Layer III.

    WMA: It stands for Windows Media Audio.

- Video file formats.

    MPEG: Motion Picture Experts Group. MPEG is an international standard multimedia file format and can compress both audio and video. MPEG-1 can be used to produce VCD files. MPEG-2 is used to compress DVD and SVCD, and support HDTV. MPEG-4 was designed for playing streaming videos.

    During the epidemic, multimedia and network played important roles in many fields. Nowadays the mobile network brings the new concept of social contact "at anytime and anywhere". Network video has come into our daily life. It mainly includes online video, on-demand TV programs, video chat, network video meetings and so on.

Word Bank

combination /ˌkɒmbɪˈneɪʃ(ə)n/ n. 混合

interchange /ˈɪntətʃeɪndʒ/ n. 互换

streaming /ˈstriːmɪŋ/ adj. 流式

contact /ˈkɒntækt/ n. 联系

on-demand 按需

text /tekst/ n. 文本

motion /ˈməʊʃn/ n. 移动，运动

concept /ˈkɒnsept/ n. 概念

demand /dɪˈmɑːnd/ n. 需要

play an important role 起着重要作用

Decide whether the following statements are true(T) or false(F) according to the text.

1. Multimedia is the combination of text and graphics. (   )
2. Animation refers to still (静止的) graphics images. (   )
3. Sound, which is also called video, is referring to sound or to things which can be heard. (   )
4. GIF can store 16 million colors. (   )

Match each of the terms to the phrases and definition that is most closely related.

A. multimedia   1. audio and video's compression standard
B. audio   2. Graphic Interchange Format
C. video   3. combination of sound, graphics, animation, video and text
D. MPEG   4. Joint Photographic Experts Group
E. JPEG   5. sound or things which can be heard
F. GIF   6. images portrayed in a television

Choose the suitable application of network video to finish the sentences.

| A. online video | B. on-demand TV programs |
| C. video chat | D. network video meeting |

1. Li Mei misses her friend in Beijing very much, so she uses _____ to talk with him.

2. In the _____, they discuss the company's development strategy.

3. Wang Yang missed a live telecast of Beijing 2022. Winter Olympic Games, so he can use _____ to watch the game.

4. Lucy often uses _____ to see films at home.

Lesson Twenty-seven

Flash memory is a form of electronic programmable memory which stores digital multimedia data immediately even when the power is switched off, such as storage cards and USB flash drives. It is often used as a storage medium in small digital products such as digital cameras, MP3 and so on.

There are kinds of flash memory in our daily life.

- CF Card (Compact Flash)

- SM (Smart Media)

- SD Card (Secure Digital Memory Card)

- MMC (Multi Media Card)

- USB (USB flash disk)

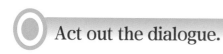 Act out the dialogue.

Linda: Hi, Zixuan. I will enter for the Beautiful Campus Video Competition. How can I make my video?

Wang Zixuan: I think you can try a most popular video editing application. It's very convenient for you to make your own videos.

Linda: So how can I use it?

Wang Zixuan: First, you should install it on your phone or computer. Then add your videos and photos. After that, you can clip the video length, add background music and even special effects. Finally, export the video in mp4 or other formats.

Linda: Sounds great, then I can store it in my new USB to share it with my friends as well. I'll go and try, thank you so much.

Wang Zixuan: You're welcome.

Word Bank

compact /kəmˈpækt/ adj. 紧密的
application /ˌæplɪˈkeɪʃ(ə)n/ n. 应用程序
effect /ɪˈfekt/ n. 效果
secure /sɪˈkjʊə(r)/ adj. 安全的
programmable /prəʊˈɡræməbl/ adj. 可编程序的

competition /ˌkɒmpəˈtɪʃ(ə)n/ n. 比赛
clip /klɪp/ v. 修剪
export /ɪkˈspɔːt/ v. 输出
storage /ˈstɔːrɪdʒ/ n. 存储器，存储
switch /swɪtʃ/ n. 开关,转换；v. 转变；（使）改变

Choose the words and fill in the blanks.

| applications | switch | competition |
| electronic | effects | |

1. How many _____ are there in your phone?
2. You have to save the documents before you _____ off the computer.
3. Linda won the first prize in English speech _____.
4. The _____ dictionary includes some videos.
5. Digital special _____ in this film are amazing.

Give your own opinions.

1. What is Flash memory?
   _____

2. What do SD Card and USB stand for?
   _____

Lesson Twenty-eight

Write the names of the devices.

Example: webcam（摄像头）

1. _____ (　)
2. _____ (　)
3. _____ (　)
4. _____ (　)

5.  _____ (   )

Learn how to make a video.

| functions   special effects   menu bar   settings   formats |

1. Start a video editing APP. Click "+" from the bottom _____ .

2. Choose materials from your photo album or shoot them now within funny _____. More videos and photos shot by digital cameras or camcorders can be transferred to your phone.

3. Clip the video length. Use more _____ from the bottom menu bar if necessary.

4. Choose music from the top and more effects and functions from the right side menu bar. For example, click the " 文 " button to add words and edit the words with colours and _____. Click the "✦" button to add more beautiful effects and try more effects from the bottom menu bar.

5. Click "the next step", select a cover and write a brief introduction for your video. Choose your location and other _____ if you like. Post it at last.

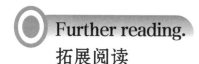

Further reading.
拓展阅读

We Media: promote rural revitalization

Self media refers to the way in which the general public releases their own facts and news to the outside world through channels such as the Internet. Self media, also known as "We Media" in English.

The platforms of the website, live streaming and e-commerce have gradually provided a new way for China to alleviate poverty（脱贫）. As a new type of live-streaming e-commerce platform, We Media is gradually integrating into the daily work of ordinary Chinese in rural areas in a more people-friendly way, becoming their "new farm tools" and a new force in the fight against poverty.

▼ **Discuss:**

1. What benefits does We Media bring to us?
2. What should we do to develop We Media or technology to make our life better?

Hints: live stream, bring convenience to sb., scientific research

Unit 8 Creating a Website

Lesson Twenty-nine

○ **Match the pictures with the software names.**

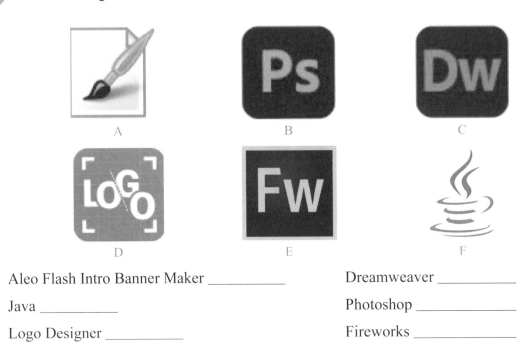

Aleo Flash Intro Banner Maker _____   Dreamweaver _____

Java _____   Photoshop _____

Logo Designer _____   Fireworks _____

○ **Choose the above software to finish the following tasks.**

1. Make the homepage of my website. _____
2. Design a logo of my website. _____
3. Design the banner of the website and an ad with words or pictures moving. _____

4. Deal with the photos of myself and try to make me look better in the photo. _____
5. The pictures for the Web are so large. I need to make them small. _____
6. The extension name of my webpage is ".jsp" _____

 Listen and fill in the blanks.

| build | design | tools | uploading |
| host | pages | tasks | create |

Dreamweaver is a web design software. It belongs to the Adobe family.

Dreamweaver helps you _____ (建设) better websites. It includes the editing _____ (工具), data and publishing tools needed to create dynamic and static Web _____ (页).

Dreamweaver includes _____ (设计) and layout tools to help you work faster and deal with major site building _____ (任务), from creating HTML, CSS, and JavaScript, to managing files and links, and _____ (上传) pages to the website.

With Dreamweaver, you can move sites easily between local and remote _____ (主机) and publish in both directions. With the help of other tools such as Photoshop, Flash, Fireworks and so on, you will _____ (创建) any Web pages you can dream up.

Word Bank

homepage /ˈhəʊmpeɪdʒ/ n. 主页
banner /ˈbænə(r)/ n. 旗帜，横幅；标语
code /kəʊd/ n. 代码
static /ˈstætɪk/ adj. 静态的
site /saɪt/ n. 网站，站点
link /lɪŋk/ n. 链接
local /ˈləʊk(ə)l/ adj. 局部的，本地的
publish /ˈpʌblɪʃ/ n. 发布
belong to 属于

logo /ˈləʊɡəʊ/ n. 标志
advertisement /ədˈvɜːtɪsmənt/ n. 广告（ad）
dynamic /daɪˈnæmɪk/ adj. 动态的
layout /ˈleɪaʊt/ n. 布局
remote /rɪˈməʊt/ adj. 远程的
upload /ˌʌpˈləʊd/ v. 上传
host /həʊst/ n. 主机
direction /dəˈrekʃ(ə)n/ n. 方向
dream up 设计，创造

Lesson Thirty

Discussion.

1. Do you know what a domain name is?
2. What does a Web host do?
3. How can we make our website more attractive?

Reading.

Do you want to have a website of your own? Follow the steps and you may create a nice website.

Step 1: Register a domain name.

A domain name is the address of your website and it looks like this: http://www.××××××.com.

Before you register your domain name, choose one which is short and easy to remember, and then type in the name to see if it's taken.

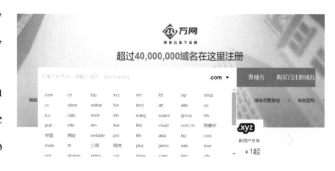

Once you have found the perfect domain name, you can fill in the form and send all the required information and the registration fee to the registry. Now you've got your domain name.

Step 2: Apply for a Web host.

A "host" is a big computer which stores the pages so that other people can see them. This computer is normally owned by special Internet companies. There are hundreds of Web hosts available. You can go to official websites to apply for a Web host.

Step 3: Make your Web pages.

Generally speaking, people use website building software such as Adobe Dreamweaver, phpDesigner, Microsoft Visual Studio and so on to design Web pages.

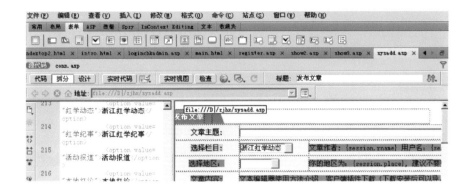

No matter what software you use, pay attention to the following useful tips on how to design it.

- "Content is the King". Fresh and unique content makes your website more attractive.
- Navigation. Make your navigation clear and available.
- Exchange links with other websites.

Step 4: Upload your pages.

When you finish all your web pages, you can use FTP tools to move them to the Web host or just use Dreamweaver uploading tools to copy the files to the remote host. Then when someone types in your domain name in the Web browser, your website will appear before his eyes. How exciting it will be!

Word Bank

register /ˈredʒɪstə(r)/ v. 注册，登记
perfect /ˈpɜːfɪkt/ adj. 完美的
registration /ˌredʒɪˈstreɪʃ(ə)n/ n. 注册，登记
own /əʊn/ n. 自己的
available /əˈveɪləb(ə)l/ adj. 有效的，可得的
tip /tɪp/ n. 小窍门，提示
fresh /freʃ/ adj. 新鲜的；新的
attractive /əˈtræktɪv/ adj. 吸引人的
apply for 申请

domain /dəˈmeɪn/ n. 域
require /rɪˈkwaɪə(r)/ v. 要求
registry /ˈredʒɪstrɪ/ n. 注册机构
special /ˈspeʃ(ə)l/ adj. 特别的；专门的
generally /ˈdʒen(ə)rəlɪ/ adv. 通常，一般地
content /ˈkɒntent/ n. 内容
unique /juːˈniːk/ adj. 独特的
exchange /ɪksˈtʃeɪndʒ/ v. 交换

Decide whether the following statements are True(T) or False(F) according to the passage.

1. A domain name is the web address that we type in the web browser. (　)
2. Some domain names are the same, so it's difficult to find your page. (　)
3. Domain names should be easy to remember. (　)
4. A host is a computer that stores the web pages. (　)
5. Dreamweaver is the only software to create websites. (　)
6. Exchanging links with other websites will bring you more visitors. (　)

Complete the sentences with the proper forms of the words or phrases given.

| the same as | navigation | exchange | hundreds of |
| apply for | generally speaking | pay attention to | own |

1. _____, it's warmer in the south than in the north.
2. Can you find a domain name which is _____ yours?
3. Many host computers are _____ by Internet companies.
4. The clear _____ can help visitors to look for information quickly.

5. When making web pages, _____ the looks of your homepage.
6. Where did you _____ the Web host?
7. Would you like to _____ links with me?
8. This is a big network with _____ computers working regularly.

Translate the following sentences into Chinese.

1. A website address is unique, just as your personal ID.
_____

2. A host is a big computer which stores your pages so that other people can see them.
_____

3. There is a saying: "Content is the King".
_____

4. When you finish all your web pages, you can use FTP tools to move them to the Web host.
_____

5. When someone types in your domain name in the Web browser, your website will appear before his eyes.
_____

Lesson Thirty-one

Read and Say.

- Dreamweaver is an Adobe's development tool to create a website, one of the most popular and easy-to-use programs.

- Java makes the global standard for developing mobile applications, games, web-based contents, and enterprise software.

- phpDesigner 8 is one of the popular PHP Web development software to help build websites, which can run in different operating systems.

- Visual Studio is a useful and powerful suite of tools to create applications.

 Act out the dialogue.

Li Lei: Your website looks very cool! How did you make it?
Linda: I used a web development software called Adobe Dreamweaver 2022.
Li Lei: Oh, what is it like?
Linda: It is a professional HTML editor for designing, coding, and developing websites, web pages and web applications.
Li Lei: I see. I think I should learn it. By the way, can you tell me how to use it?
Linda: Sure. Let me show you how to do it. It's easy.
Li Lei: Great!

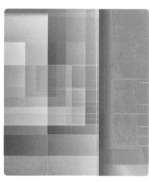

 Do you know the workspace of Dreamweaver 2022? Find their names.

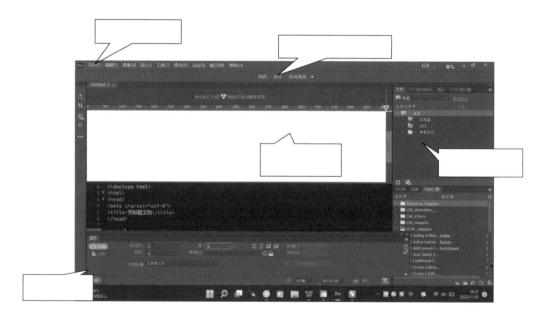

A. Menu Bar（菜单栏） B. Panel Groups（面板组）
C. Property Inspector（属性检查器） D. Document Window（文档窗口）
E. Document Toolbar（文档工具栏）

Can you find out the software that they usually use?

Mr. Wang is a computer engineer. He designs mobile phone games. He may use _____.

Mary is a computer learner. She wants to create her own website but she knows little about coding languages. She just makes some static web pages. She will use _____.

Lin Tao works in a computer company. He creates different applications for some enterprises. He usually uses _____ _____.

Miss Yang is good at building websites. She often designs websites that can run in UNIX or Linux as well as Windows. She often uses _____

Sam runs a dynamic website. He usually uses _____ to make web pages.

Word Bank

creation /krɪˈeɪʃ(ə)n/ n. 创造
enterprise /ˈentəpraɪz/ n. 企（事）业单位

base /beɪs/ v. 以……为基础
editor /ˈedɪtə(r)/ n. [计] 编辑器，编辑程序

Lesson Thirty-two

Do you know the meaning of the following network terms? Try to match them.

1. Domain Name
2. Web Browser
3. LAN (Local Area Network)
4. WAN (Wide Area Network)
5. Home Page
6. Search Engine
7. HTTP（Hypertext Transfer Protocol）

A. 广域网
B. Web 浏览器
C. 局域网
D. 域名
E. 搜索引擎
F. 超文本传输协议
G. 主页

Have you ever seen such information when visiting a website? Match the pictures with their English descriptions.

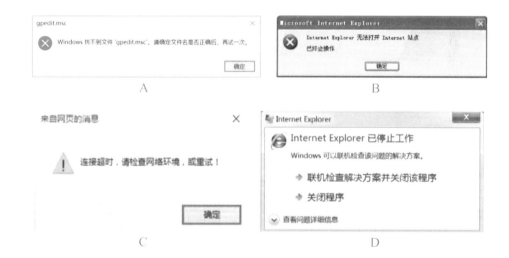

E

F

_____ 1. A connection with the server could not be established.

_____ 2. Internet Explorer cannot open the Internet site.

_____ 3. Internet Explorer has encountered a problem and needs to close.

_____ 4. The operation timed out.

_____ 5. The system cannot find the file specified.

_____ 6. The plug-in has performed an illegal operation.

Do you know how to upload your Web pages to your Web host? Find the right description for each step.

Step 1 _____

Step 2_____

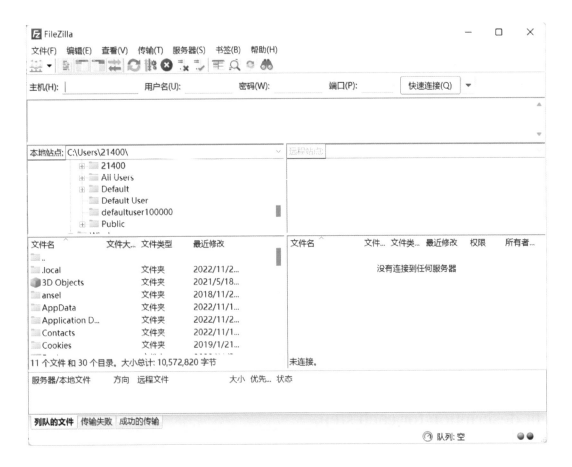

Step 3_____

A. Run the program FileZilla. Enter your username, password, and FTP address beginning with "ftp", then click "Quick Link" button.

B. Now you've entered your Web host. Choose your files in the local computer, then copy them to the Web host.

C. Download and install FileZilla client, a free program that provides FTP upload service.

Do you know the following words? Please match the English words with the right Chinese meanings.

one-key dress-up   保存
template   导航栏
website   贴图
double clicking   装扮商城
navigation   双击
stickers   模板
dress-up mall   一键装扮
save   网站

 Do you know how to dress up your QQ zone? Fill in the blanks.

| one-key dress-up | template | website | double clicking |
| navigation | stickers | dress-up mall | save |

Step 1. Enter QQ zone from QQ panel（模板）or the _____ of QQ.

Step 2. You can use _____. It's fast and easy. Choose one of the styles or themes（主题）, and then you've got it done.

Step 3. If you want to make more variations（变化）, click the _____ button（按钮）, and enter advanced settings（高级设置）.

Step 4. You can change the title（标题）of your QQ Zone by _____ it.

Step 5. Choose your favorite _____. Some of them are free, others will cost you some money.

Step 6. You can also try DIY dress-up. In this part you can change the skin, animation（动画）, title bar（标题栏）and _____ bar.

Step 7. You can choose template or _____ to change your background（背景）.

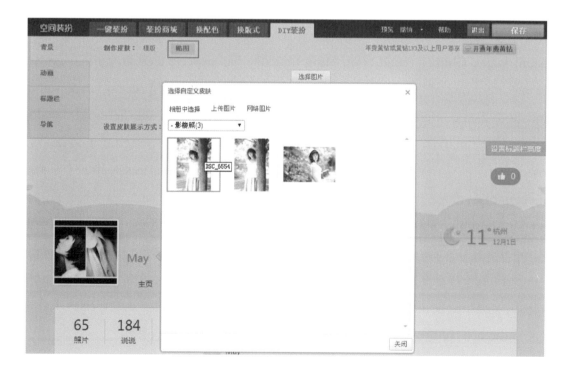

Step 8. You can preview(预览)your new QQ zone after dressing it up. If you are pleased with the new change, then _____ it.

Further reading.
拓展阅读

In recent years, China's cloud computing industry has experienced an annual growth rate of over 30%, making it one of the fastest growing markets in the world. In recent years, remote office, online education, web conference and other needs have further promoted the rapid development of the cloud computing market. Cloud computing is gradually becoming a digital innovation platform that empowers(给予……权力)the digital economy, becoming the infrastructure(基础设施)of the digital economy.

Cloud Site-building Service is a method based on cloud technology. It has more advantages than traditional ways, such as diversified(多样化的)station building functions, standardized services, integrated(集成的)cloud suite resources, security protection, low cost and fast self-service website construction and so on. It does provide convenience for enterprises(企业). We believe that in the near future, China's cloud computing will lead the world!

▼ **Discuss:**

1. What are the functions of Cloud Site-building Service?
2. What will Chinese Cloud Computing be like in the future?

Hints: save costs, improve efficiency, provide personalized services

Unit 9   Network Security

Lesson Thirty-three

Match the icons with the antivirus software.

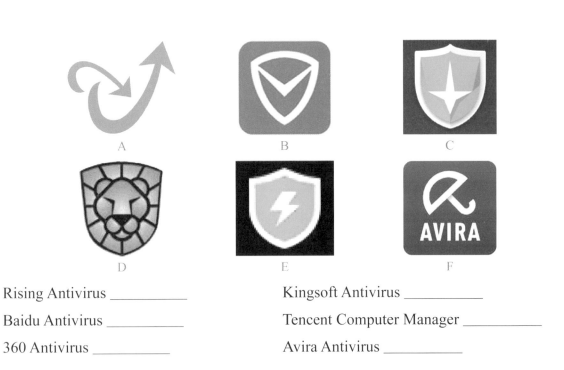

Rising Antivirus _____   Kingsoft Antivirus _____

Baidu Antivirus _____   Tencent Computer Manager _____

360 Antivirus _____   Avira Antivirus _____

Find out the correct names in the Antivirus software interface.

Quick Scan   Full Scan   Custom Scan

快速扫描 _____

全盘扫描 _____

自定义扫描 _____

 Listen and fill in the blanks.

antivirus   installed   security   threats

1. The world's first _____ (杀毒) software was born in 1989.
2. It is a type of software used to eliminate computer _____ (威胁) such as viruses and trojans.
3. A computer does not need to have two or more antivirus software _____ (被安装) at the same time under each operating system.
4. The "Cloud Security" is the latest manifestation of information _____ (安全) in the network era.

Listen and decide if the statements are True(T) or False(F).

(　) 1. A spyware is a kind of antivirus software.
(　) 2. Spyware collects the information of your computer and sends it to another computer.
(　) 3. A spywares is sometimes unknowingly downloaded and installed by users.
(　) 4. There is only one tool to deal with spyware, Spybot-Search & Destroy.
(　) 5. Spybot-Search & Destroy can scan computers and delete any or all the spyware.

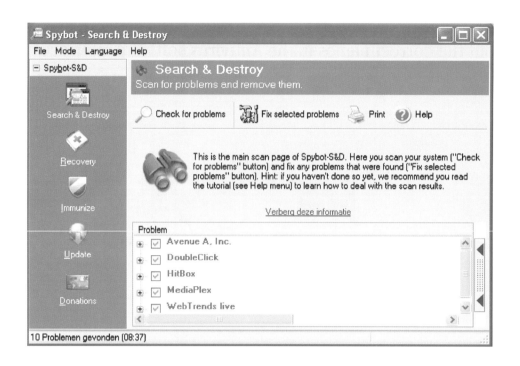

Word Bank

antivirus /ˈæntɪvaɪrəs/ n. 防病毒
research /rɪˈsɜːtʃ/ v. 研究
spyware /ˈspaɪweə(r)/ n. 间谍软件
unknowingly /ʌnˈnəʊɪŋli/ adv. 不知不觉地

professional /prəˈfeʃən(ə)l/ adj. 专业的
successfully /səkˈsesfəli/ adv. 成功地
collect /kəˈlekt/ v. 收集

Lesson Thirty-four

 Discussion

1. What is a computer virus?
2. What effects can viruses make on the computers?
3. What should we do to protect our computer?

Reading.

Protect Your Computer from Viruses

What is a computer virus? It is a program designed to spread from one computer to another and to interfere with computer operation.

A virus might destroy or delete data on your computer, use your email program to spread itself to other computers, or even erase everything on your computer.

Computer viruses are often spread through downloads on the Internet or by attachments in email messages.

Here are some common tips that help you protect your computer from viruses:

1. Do not install software that you do not trust.
2. Make sure you upgrade your firewall and antivirus software on time.
3. Immediately delete emails from people that you do not know.

4. If you must download and install a software, make sure it is from a safe website.
5. Do not go to websites that you are unsure of or never hear of.
6. Do not click on those advertisements offering you a free prize or something else.
7. Keep a habit of virus scanning once a week or even once a day.
8. Always back up your files in case they are destroyed or deleted.

Word Bank

spread /spred/ *v.* 传播
destroy /dɪ'strɔɪ/ *v.* 破坏
attachment /ə'tætʃmənt/ *n.* 附件
protect /prə'tekt/ *v.* 保护
trust /trʌst/ *v.* 相信，信任
upgrade /ˌʌp'ɡreɪd/ *v.* 升级
prize /praɪz/ *n.* 奖赏
common sense 常识

interfere /ˌɪntə'fɪə(r)/ *v.* 干扰
erase /ɪ'reɪz/ *v.* 抹去；擦除
disguise /dɪs'ɡaɪz/ *v.* 伪装
install /ɪn'stɔːl/ *v.* 安装
intend /ɪn'tend/ *v.* 想要；打算
firewall /'faɪəwɔːl/ *n.* 防火墙
keep a habit of 保持……习惯

 Decide Whether the following statements are True(T) or False(F).

(   ) 1. A computer virus is a kind of program.
(   ) 2. A virus can make a computer not work properly, or even break down.
(   ) 3. If a virus gets into your computer, you don't have to remove it until your computer gets in trouble.
(   ) 4. The firewall can stop virus from getting into your computer.
(   ) 5. Upgrading the antivirus software is good to our computer.

 Fill in the blanks with the proper forms of the words or expressions given.

| upgrade | install | back up | scan | delete | download |

1. Do not _____ software that you do not trust.
2. Only _____ and install a software that is from a safe website.

3. _____ the emails from people that you don't know.
4. _____ your files once a week or even once a day.
5. Always _____ your file in case they are destroyed or deleted.
6. _____ your antivirus and firewall software on time.

Lesson Thirty-five

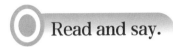 Read and say.

- Trojan is a virus designed to cause some harmful activities or to provide a backdoor to your system.

- Worm is a virus hiding in active memory and duplicate itself to make the computer run slowly.

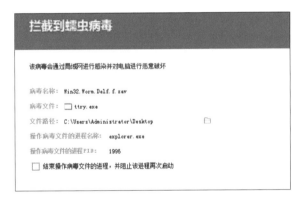

- Spyware is a program that monitors your activity or information on your computer and sends information to a remote computer unknowingly.

- Macro virus is a virus which infects Office files such as Word, Excel, PowerPoint documents. It spreads through Office templates.

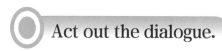 Act out the dialogue.

Tom: Oh, my God! The computer froze again! That's the third time today! Hey Li Lei can you come and take a look at my PC? It's acting up again. It must have a virus or something.

Li Lei: Just give me a second. I'll be right up.

Tom: OK, but be quick.

(A few minutes later)

Li Lei: So I ran a virus scan on your computer, and it turns out that you have a lot of infected files!

Tom: But I'm quite careful when I'm browsing the Internet. I have no idea how I could have picked up a virus.

Li Lei: Well, you have to make sure that your antivirus software is updated regularly. Yours wasn't the up-to-date version, and that's probably what was causing your problems.

Tom: OK.

 Can you find the right picture for each virus?

(　　) System virus begins with "Win32" "PE" "Win95", etc. It often infects Windows system files.

(　　) Worm virus begins with "Worm". It spreads through the Internet or system leaks.

(　　) Trojan virus begins with "Trojan". It get into your computer through the Internet or system leaks and give away your information.

(　　) Macro virus begins with "Macro". It is one of script viruses. It infects the Office documents.

(　　) Backdoor virus begins with "Backdoor". It spreads through the Internet and allows your computer to be controlled by remote users.

(　　) Joke virus begins with "Joke". It usually frightens computer users, but it won't destroy computer files.

A.

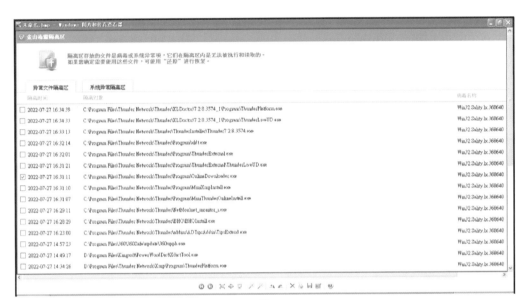

B.

C.

D.

E.

F.

Do you know which of the following problems might be caused by viruses?

- ☐ My chatting software number is stolen.
- ☐ The power doesn't work
- ☐ The computer restarts again and again.
- ☐ The computer runs too slowly.
- ☐ My computer often pop-up ads.
- ☐ My computer can't connect to the Internet.
- ☐ The software edition is low.
- ☐ There is no signal on the screen.
- ☐ Many files are lost.
- ☐ The keyboard doesn't work.
- ☐ The fan doesn't run any more.

Word Bank

duplicate /ˈdjuːplɪkeɪt/ v. 复制；使加倍
freeze /friːz/ v. 冻结；死机
act up 运作不正常；出毛病
macro /ˈmækrəʊ/ n. 宏
infected /ɪnˈfektɪd/ adj. 被感染的

Lesson Thirty-six

Match the English with the Chinese meaning.

1. virus	A. 宏病毒
2. antivirus software	B. 病毒
3. update	C. 蠕虫病毒
4. firewall	D. 防火墙
5. password	E. 防毒软件
6. worm	F. 木马病毒
7. Trojan	G. 密码
8. up-to-date	H. 最新的
9. infect	I. 更新
10. macro viruses	J. 感染

Translate the following words or phrases.

1. 扫描 _____
2. 更新 _____
3. 升级 _____
4. 弹出广告 _____
5. firewall _____
6. common sense _____
7. antivirus software _____
8. network security _____

Complete the following sentences.

1. I'm afraid your computer _____（感染病毒了）.
2. Why not _____（升级你的杀毒软件呢）?
3. I am afraid _____（版本太低了）.
4. Can you tell me how to _____ my computer _____ virus（保护……免遭）?
5. A computer virus _____（运行）in memory and _____（存储）in a file.

Learn what to do when your computer is infected by viruses. Choose the title for each picture.

> A. Enter the Safe Mode.
> B. Decide whether your computer is infected by virus.
> C. Restart your computer.
> D. Run antivirus software.

Step 1 _____

The computer is as slow as a snail.
CPU usage is high.
Many unknown processes are running.
…

Step 2 _____

Reboot, press F8, and enter boot option.
Choose the Safe Mode.

Step 3 _____

Use the antivirus software to scan and kill the viruses.
Delete the infected files manually if they can't be deleted automatically.

Step 4 _____

Restart your computer.
If that fails, call an expert for help or restore the system.

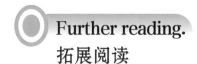

Further reading.
拓展阅读

China's Approach to Cyberspace Governance

With the digital age having a profound impact on human society, China has put forward "Chinese solutions" for the international cyber governance system (国际网络空间治理体系), demonstrated by its efforts in maintaining cyber security, developing global internet infrastructure and building an internet governance system, etc.

▼ **What kind of information do you need to protect when using the Internet?**

☐ password ☐ your name and address ☐ your nickname
☐ your bank account ☐ others _____

▼ **Discuss:**

How to protect your private information while surfing the Internet?

Hints: set complicated passwords (设定复杂的密码), avoid using your real name or spelling, screenshots, Wi-Fi, social media

Unit 10　Computer Maintenance

Lesson Thirty-seven

Match the Chinese meanings with the English items.

```
              ROM PCI/ISA BIOS (2A69KG0D)
                 CMOS SETUP UTILITY
                AWARD SOFTWARE, INC.

 STANDARD CMOS SETUP            INTEGRATED PERIPHERALS
 BIOS FEATURES SETUP            SUPERVISOR PASSWORD
 CHIPSET FEATURES SETUP         USER PASSWORD
 POWER MANAGEMENT SETUP         IDE HDD AUTO DETECTION
 PNP/PCI CONFIGURATION          SAVE & EXIT SETUP
 LOAD BIOS DEFAULTS             EXIT WITHOUT SAVING
 LOAD PERFORMANCE DEFAULTS

 Esc : Quit                     ↑↓→← : Select Item
 F10 : Save & Exit Setup        (Shift) F2 : Change Color

              Time, Date, Hard Disk Type...
```

用户密码 _____	集成外围设备 _____
BIOS 功能设置 _____	不保存退出 _____
载入 BIOS 默认值 _____	芯片组功能设置 _____
IDE 硬盘自动检测 _____	超级用户密码 _____
电源管理设置 _____	标准 CMOS 设置 _____
保存并退出 _____	即插即用 /PCI 配置 _____
载入高性能默认值 _____	

Listen to the dialogue and reorder the following sentences.

(1) Hello, is that HP Tech Support?

(   ) Sounds like a virus.

(   ) What should I do? Should I reinstall the system?

(   ) Yes, can I help you?

(   ) I don't know the details of your situation. Reinstalling is an option, or you can install some antivirus software.

(   ) Something has gone wrong with my computer. It always restarts.

(7) Thanks a lot.

 Learn how to clean the screen of your computer.

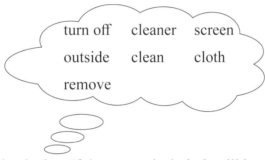

Step 1. _____（关闭）the device. If the screen is dark, it will be easier to see the dirty area.

Step 2. Use a dry, soft _____（布）and wipe the screen gently. If the dry cloth does not completely _____（除掉）the dirt or oil, do not press harder or you will make pixels burn out.

Step 3. Spray some screen _____（清洁剂）or 50:50 vinegar and water on the screen. Then wipe it from the inside to the _____（外面）.

Step 4. _____（擦除）the dirt on the plastic edge that surrounds the screen with any multipurpose cleaner but take care not to contact the _____（屏幕）itself.

Word Bank

reinstall /ˌriːɪnˈstɔːl/ v. 重新安装

situation /ˌsɪtʃuˈeɪʃ(ə)n/ n. 形势；情况

dirty /ˈdɜːti/ adj. 脏的

cloth /klɒθ/ n. 布；织物

detail /ˈdiːteɪl/ n. 详情

option /ˈɒpʃ(ə)n/ n. 选择，选项

area /ˈeərɪə/ n. 地区；范围

wipe /waɪp/ v. 揩，擦

gently /'dʒentli/ adv. 温柔地；轻轻地
oil /ɔɪl/ n. 油；石油
spray /spreɪ/ v. 喷洒
plastic /'plæstɪk/ adj. 塑料的
surround /sə'raʊnd/ v. 围绕，包围
contact /'kɒntækt/ v. 联系

completely /kəm'pli:tli/ adv. 完全地，彻底地
pixel /'pɪksl/ n. 像素
vinegar /'vɪnɪɡə(r)/ n. 醋
edge /edʒ/ n. 边缘
multipurpose /ˌmʌltɪ'pɜ:pəs/ adj. 多用途的

Lesson Thirty-eight

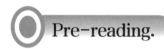
Pre-reading.

1. What problems have you met with when you use your computer?
2. How do you often solve your computer problems?
3. What will cause your computer to break down?

Reading.

扫一扫
听录音

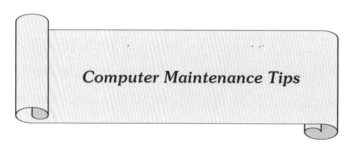
Computer Maintenance Tips

Do you know how to take care of your computer? Now it is a good time to form some good computer maintenance habits before you get into trouble.

1. Never turn your computer off with the power switch until Windows has shut down.

When your computer locks up and your hard drive is not running, you can turn the power off without harmful effects to the hard drive. As cutting the power can also result in losing data or Windows files, you should only do this when you have to.

2. Backup important files.

You must remember to store your important files at different drives, especially not at drive C.

You can backup data to U-disk, removable disks, CDs, etc. The time to backup data is when you create something important to you. Don't wait until tomorrow.

3. Run Scandisk at least once a month and antivirus software every day.

This will keep your hard drive healthy and prevent crashes. Install one of the antivirus software. And run the antivirus software every day.

4. Do not let a lot of programs load up when you start your computer.

Some programs will load up when you start your computer. All programs in your Windows System Tray (in the lower right of your screen) are running on your computer. Close them if you don't need them and use antivirus software to stop them loading when your computer boots up.

5. Keep the screen and the keyboard clean.

Clean the screen and the keyboard as well as the mouse regularly with a piece of soft cloth and special cleanser. Don't drink or eat while using your computer, or you will easily get them dirty.

I hope these computer maintenance tips will keep you out of trouble. However, if you still have problems with your computer, call an expert to help you.

Word Bank

maintenance /ˈmeɪntənəns/ *n.* 维护
switch /swɪtʃ/ *n.* 开关
scandisk /ˈskændɪsk/ *n.* 磁盘检查工具
cleanser /ˈklenzə(r)/ *n.* 清洁剂

separate /ˈseprət/ *adj.* 分开的，单独的
crash /kræʃ/ *n.* 碰撞，坠落，崩溃
tray /treɪ/ *n.* 盘，碟，托盘
configure /kənˈfɪɡə(r)/ *v.* 配置

Choose the best answers according to the passage.

1. You can't turn off the power when _____.
   A. the disk isn't running.   B. the disk light is blinking（闪烁）.
   C. we leave the office.

2. If the computer shuts down suddenly, what may NOT happen?
   A. Data will be lost.   B. Windows files will be missing.
   C. Keyboard doesn't work.

3. If my computer doesn't have enough memory, what shall I do?
   A. Uninstall some programs.   B. Install some software.
   C. Run the antivirus software.

4. When we use the computer, what shouldn't we do at the same time?
   A. Drink or eat.   B. Listen to music.
   C. Chat with friends.

Fill in the blanks according to the passage.

1. Getting some knowledge of computer maintenance may help you out of t_____.
2. You can't cut the power until your Windows has s_____ down.
3. Scandisk can keep your drives h_____ and prevent from crashes.

4. We should keep the screen, keyboard and mouse c _____.
5. You can ask an e _____ for help if you fail to solve the problem.

Lesson Thirty-nine

Read and say.

If your computer doesn't boot, you will hear different beeps. In an Award BIOS motherboard, different beeps means different trouble.

* 1 long 1 short: Memory or motherboard is faulty.
* 1 long 2 short: Graphics card is not into place, or faulty.
* 1 long 3 short: Keyboard error.
* 1 long 9 short: Motherboard flash memory error.
* 2 short: CMOS setup error.
* Continuous long: No memory, or memory is not into place or could be faulty.
* Continuous short: Power error.

Act out the dialogue.

Linda: My computer is so slow. I can't stand it.
Tom: Let me see. Oh, too many programs are running in the background.
Linda: What should I do then?
Tom: You must shut down some background programs.
Linda: Can you show me how to do it?
Tom: Sure. Press Ctrl + Alt + Del keys and come into the Windows Task Manager.
Linda: OK. And then?
Tom: Choose the background programs and click the **End the Task** button.
Linda: OK. It's done. I see. It's a little faster than before. Thank you very much.
Tom: You are welcome.

Word Bank

beep /biːp/ *n.* 蜂鸣声　　　　　　　　　faulty /ˈfɔːlti/ *adj.* 有错误的

error /ˈerə(r)/ *n.* 错误　　　　　　　　　stand /stænd/ *v.* 忍受

Task Manager 任务管理器

Do you know the following English words? Please match the words with the right Chinese meanings.

motherboard	错误
memory	内存
error	显卡
graphics card	后台程序
background programs	结束进程
End the Task	主板
stand	忍受

My computer usually works very well. But today it's acting up. Can you help me with the troubleshooting?

1. The computer doesn't boot up. Two short beeps are heard.　　　　Solution: (　　)
2. There is no display on the screen.　　　　　　　　　　　　　　Solution: (　　)
3. The computer can't enter OS. It shows NTLDR is missing.　　　　Solution: (　　)
4. The computer can't connect to the Internet.　　　　　　　　　　Solution: (　　)
5. I can't open QQ. It shows the program setting is incorrect.　　　　Solution: (　　)

> A. Open the case（机箱）, remove（去除）the dirt and dusk（灰尘）on the memory and the motherboard.
> B. Uninstall（卸载）the software and then reinstall（重新安装）it again.
> C. When the power is switched on（打开）, hold down the Del key and enter CMOS setup utility. Choose LOAD BIOS DEFAULTS. Then save and exit（退出）.
> D. That means a system file is missing. You have to reinstall the system.

E. Make sure the IP address is correct（正确）. If so, disable（禁用）the network connection and then enable（启用）it again.

Lesson Forty

Find the antonyms of the following words.

start	A. escape
enter	B. power off
install	C. clean
power on	D. uninstall
dirty	E. legal
visual	F. output
valid	G. invalid
illegal	H. invisible
download	I. end
input	J. upload

What can we do to maintain our computer? Fill in the blanks.

1. We should keep the _____（屏幕和键盘）clean.

2. We shouldn't _____（关闭电源）our computer until Windows has shut down.

3. We should _____（备份）important files to different drives.

4. We should run _____（磁盘检查工具）once a month and the antivirus software every day.

5. We shouldn't let unnecessary programs _____（启动）so that the computer can run a little fast.

6. Uninstall the software that you don't use and then _____（重启）the computer.

7. When a system file is missing, we have to _____（重新安装）the system.

8. When there is too much dirt on the motherboard, we should _____（清除）it before the computer goes wrong.

Do you know how to create a new partition in Windows system? Reorder the steps.

1

Step _____

You will see the window of *Computer Management*, find and click the item of *Disk Management* on the left column（栏）.

2

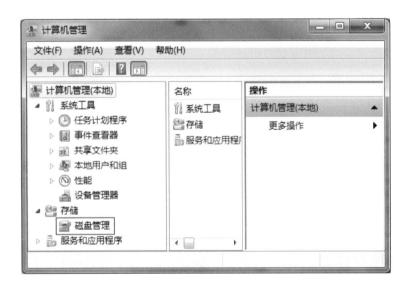

Step _____

Click *Start* and right click on *Computer*. Then click on *Management*.

3

Step _____

Right click on free space which is usually in green and click on *New logic drive*.

4

Step _____

Enter the partition size. The default is to set all free space as one drive. If you set it two or more drives, enter a proper number.

5

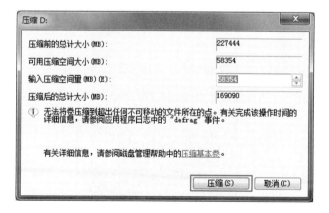

Step _____

Choose *Compress volume*. The drive will be formatted. It can't be used until the formatting has been finished.

Further reading.
拓展阅读

Intellectual Manufacturing Delighted the 2022 Beijing Winter Olympics: From Made in China to Created in China

The Beijing Winter Olympics is unprecedented in the application of science and technology, which will hopefully be widely used around the world. Here is an example:

In the context of 5G, artificial intelligence and the Internet of Things, the national stadium—known as the "Bird's Nest"—has undergone an intelligent transformation to host the opening and closing ceremonies. For example, digital technology has been used in equipment operation, energy management and environmental quality control, providing favorable conditions for the ceremonies.

▼ **Can you name some high-tech products used in the 2022 Beijing Winter Olympics?**

☐ special camera system ☐ smart restaurant ☐ smart bed
☐ guide robot ☐ others_____

▼ **Discuss:**

Can you talk about other high-tech products in China that has changed your life?

Hints: 5G smart phone, high-speed train, artificial intelligence, visual reality, new energy vehicle

Glossary

a wide range of …… 大范围的，各种不同的	L7（表示所在课文，下同）
access /'ækses/ *n.* 入口；访问，存取	L6
act up 运作不正常；出毛病	L35
add /ˌeɪ diː 'diː/ *v.* 添加	L11
advantage /əd'vɑːntɪdʒ/ *n.* 优点	L15
advertisement /əd'vɜːtɪsmənt/ *n.* 广告（缩写 ad）	L29
affect /ə'fekt/ *v.* 影响	L6
afford /ə'fɔːd/ *v.* 负担得起	L3
agriculture /'æɡrɪkʌltʃə(r)/ *n.* 农业	L22
agricultural /ˌæɡrɪ'kʌltʃərəl/ *adj.* 农业的	L23
aid /eɪd/ *n.* 帮助	L10
AMD（Advanced Micro Devices, Inc.）超威半导体公司	L7
analog /'ænəlɒɡ/ *adj.* 模拟的	L7
analog ICs 模拟芯片	L7
animation /ˌænɪ'meɪʃ(ə)n/ *n.* 动画	L9
antivirus /ˈæntivaɪrəs/ *n.* 防病毒	L33
application /ˌæplɪ'keɪʃ(ə)n/ *n.* 应用	L22
apply /ə'plaɪ/ *v.* 适用，应用	L2
apply for 申请	L30
archive /'ɑːkaɪv/ *n.* 档案文件；*v.* 存档	L11
archive manager 档案文件管理器	L11
area /'eəriə/ *n.* 地区；范围	L37
attachment /ə'tætʃmənt/ *n.* 附件	L34
attractive /ə'træktɪv/ *adj.* 吸引人的	L30

automatic /ˌɔːtəˈmætɪk/ adj. 自动的	L22
automobile /ˈɔːtəməbiːl/ n. 汽车	L7
available /əˈveɪləb(ə)l/ adj. 有效的，可得的	L30
back up 备份	L11
banner /ˈbænə(r)/ n. 旗帜，横幅；标语	L29
base /beɪs/ v. 以……为基础	L31
be made up of 由……组成	L13
be related to 与……有关	L14
beep /biːp/ n. 蜂鸣声	L39
BeiDou positioning function 北斗定位功能	L22
belong to 属于	L29
billion /ˈbɪljən/ n. 十亿	L22
biomedicine /ˌbaɪəʊˈmedɪsɪn/ n. 生物医学	L2
Bluetooth 蓝牙	L2
boom /buːm/ v. 迅速发展	L22
browser /ˈbraʊzə(r)/ n. 浏览器	L10
button /ˈbʌt(ə)n/ n. 按钮	L11
CAD: Computer-Aided Design 计算机辅助设计	L10
camcorder /ˈkæmkɔːdə(r)/ n. 便携式摄像机	L25
camera /ˈkæm(ə)rə/ n. 照相机	L5
capacity /kəˈpæsəti/ n. 容量	L6
carelessly /ˈkeələsli/ adv. 粗心地	L19
central /ˈsentrəl/ adj. 中心的；主要的	L6
chart /tʃɑːt/ n. 图表	L17
chip /tʃɪp/ n. 芯片	L1
choose /tʃuːz/ v. 选择	L11
circuit /ˈsɜːkɪt/ n. 电路	L2
cleanser /ˈklenzə(r)/ n. 清洁剂	L38
clip /klɪp/ v. 修剪	L27
cloth /klɒθ/ n. 布；织物	L37
Cloud Computing 云计算	L2
code /kəʊd/ n. 代码	L29
collect /kəˈlekt/ v. 收集	L33

collection /kəˈlekʃ(ə)n/ n. 集合	L19
combination /ˌkɒmbɪˈneɪʃ(ə)n/ n. 结合体	L23
command /kəˈmɑːnd/ n. 指令	L5
commerce /ˈkɒmɜːs/ n. 贸易，商业	L21
common /ˈkɒmən/ adj. 常见的	L6
common sense 常识	L34
commonly /ˈkɒmənli/ adv. 通常地	L11
communication /kəˌmjuːnɪˈkeɪʃ(ə)n/ n. 交流，通信	L14
compact /kəmˈpækt/ adj. 紧密的；小型的	L27
compatibility /kəmˌpætəˈbɪləti/ n. 兼容性	L18
competition /ˌkɒmpəˈtɪʃ(ə)n/ n. 比赛	L27
completely /kəmˈpliːtli/ adv. 完全地，彻底地	L37
complex /ˈkɒmpleks/ n. 复合体	L1
comprehensive /ˌkɒmprɪˈhensɪv/ adj. 综合的	L18
compress /kəmˈpres/ v. 压缩	L11
concept /ˈkɒnsept/ n. 概念	L26
configure /kənˈfɪɡə(r)/ v. 配置	L38
connect /kəˈnekt/ v. 连接	L14
consider /kənˈsɪdə(r)/ v. 认为	L21
contact /ˈkɒntækt/ v. 联系	L26
content /ˈkɒntent/ n. 内容	L30
contribute to 有助于	L22
control /kənˈtrəʊl/ v. 控制，管理	L10
convenient /kənˈviːniənt/ adj. 方便的	L14
convey /kənˈveɪ/ v. 传送	L18
core /kɔː(r)/ n. 核心	L7
corporation /ˌkɔːpəˈreɪʃ(ə)n/ n. 公司	L1
crash /kræʃ/ n. 碰撞，坠落，崩溃	L38
create /kriˈeɪt/ v. 创建	L19
creation /kriˈeɪʃ(ə)n/ n. 创造	L31
crop /krɒp/ v. 剪短；剪裁	L17
data /ˈdeɪtə/ n. 数据；资料	L2
DBMS: Database Management System 数据库管理系统	L10

decompress /ˌdiːkəmˈpres/ v. 解压缩	L11
default /dɪˈfɔːlt/ n. 默认	L11
degree /dɪˈɡriː/ n. 程度	L14
delete /dɪˈliːt/ v. 删除	L19
demand /dɪˈmɑːnd/ n. 需要	L23
depend on 取决于	L3
depend /dɪˈpend/ v. 依靠	L3
design /dɪˈzaɪn/ v. 设计	L2
desktop /ˈdesktɒp/ adj. 台式的	L1
destroy /dɪˈstrɔɪ/ v. 破坏	L34
detail /ˈdiːteɪl/ n. 详情	L37
development /dɪˈveləpmənt/ n. 发展，开发	L13
device /dɪˈvaɪs/ n. 装置；设备	L5
digital /ˈdɪdʒɪt(ə)l/ adj. 数字的	L17
digital ICs 数字芯片	L7
direction /dəˈrekʃ(ə)n/ n. 方向	L29
dirty /ˈdɜːti/ adj. 脏的	L37
disguise /dɪsˈɡaɪz/ v. 伪装	L34
disk /dɪsk/ n. 磁盘	L5
display /dɪˈspleɪ/ v. 显示；n. 显示器	L5
divide……into…… 把……分成……	L6
document /ˈdɒkjumənt/ n. 文档	L17
domain /dəˈmeɪn/ n. 域	L30
download /ˌdaʊnˈləʊd/ v. 下载	L15
dream up 设计，创造	L29
duplicate /ˈdjuːplɪkeɪt/ v. 复制；使加倍	L35
dynamic /daɪˈnæmɪk/ adj. 动态的	L29
economy /ɪˈkɒnəmi/ n. 经济	L22
edge /edʒ/ n. 边缘	L37
edit /ˈedɪt/ v. 编辑	L17
editor /ˈedɪtə(r)/ n. [计] 编辑器，编辑程序	L31
education /ˌedʒuˈkeɪʃ(ə)n/ n. 教育	L21
effect /ɪˈfekt/ n. 效果	L27

efficiency /ɪˈfɪʃ(ə)nsi/ n. 效率 — L18

electronic /ɪˌlekˈtrɒnɪk/ adj. 电子的，电子学的 — L7

electronics /ɪˌlekˈtrɒnɪks/ n. 电子器件 — L3

eliminate /ɪˈlɪmɪneɪt/ v. 淘汰 — L7

energy /ˈenədʒi/ n. 能源 — L2

engine /ˈendʒɪn/ n. 引擎，发动机 — L21

engineer /ˌendʒɪˈnɪə(r)/ n. 工程师 — L13

enterprise /ˈentəpraɪz/ n. 企（事）业单位 — L31

entertainment /ˌentəˈteɪnmənt/ n. 娱乐 — L21

environment /ɪnˈvaɪrənmənt/ n. 环境 — L14

equipment /ɪˈkwɪpmənt/ n. 设备 — L7

erase /ɪˈreɪz/ v. 抹去；擦除 — L34

error /ˈerə(r)/ n. 错误 — L39

exceptional /ɪkˈsepʃən(ə)l/ adj. 杰出的 — L21

exchange /ɪksˈtʃeɪndʒ/ v. 交换 — L30

expandable /ɪkˈspændəbl/ adj. 可扩展的 — L3

expert /ˈekspɜːt/ n. 专家 — L13

export /ɪkˈspɔːt; ˈekspɔːt/ v. 输出 — L27

extension /ɪkˈstenʃ(ə)n/ n. 扩展 — L19

facility /fəˈsɪləti/ n. 设施，设备 — L2

failed /feɪld/ adj. 失败的 — L14

fan /fæn/ n. 风扇 — L5

faulty /ˈfɔːlti/ adj. 有错误的 — L39

feature /ˈfiːtʃə(r)/ n. 特征，特性 — L14

feature-rich adj. 丰富的 — L18

fee /fiː/ n. 费用 — L22

field /fiːld/ n. 领域 — L2

file format 文件格式 — L11

firewall /ˈfaɪəwɔːl/ n. 防火墙 — L34

flourish /ˈflʌrɪʃ/ v. 繁荣 — L22

folder /ˈfəʊldər/ n. 文件夹 — L19

format /ˈfɔːmæt/ n. 格式 — L11

founder /ˈfaʊndə(r)/ n. 创始人 — L13

单词	课次
free /friː/ *adj.* 免费的	L15
freeze /friːz/ *v.* 冻结；死机	L35
fresh /freʃ/ *adj.* 新鲜的；新的	L30
function /'fʌŋkʃ(ə)n/ *n.* 功能	L10
gather /'gæðə(r)/ *v.* 收集	L17
generally /'dʒen(ə)rəli/ *adv.* 通常，一般地	L30
generate /'dʒenəreɪt/ *v.* 生成	L25
generation /ˌdʒenə'reɪʃ(ə)n/ *n.* 一代	L2
gently /'dʒentli/ *adv.* 温柔地；轻轻地	L37
giant /'dʒaɪənt/ *adj.* 巨大的	L2
global /'gləʊb(ə)l/ *adj.* 全球的	L22
GPU *abbr.* 图形处理器（Graphics Processing Unit）	L7
graphic /'græfɪk/ *adj.* 图解的；*n.* 图形，图表	L17
graphical /'græfɪkl/ *adj.* 图形的	L14
headphone /'hedfəʊn/ *n.* 耳机	L5
homepage /'həʊmpeɪdʒ/ *n.* 主页	L29
host /həʊst/ *n.* 主机	L29
huge /hjuːdʒ/ *adj.* 巨大的	L23
HYGON 海光	L7
icon /'aɪkɒn/ *n.* 图标	L6
improve /ɪm'pruːv/ *v.* 改进	L18
in general 一般而言；通常	L3
include /ɪn'kluːd/ *v.* 包含，包括	L10
including /ɪn'kluːdɪŋ/ *prep.* 包含，包括	L10
incorporated /ɪn'kɔːpəreɪtɪd/ *adj.* 股份有限的	L13
industry /'ɪndəstri/ *n.* 工业	L22
infected /ɪn'fektɪd/ *adj.* 被感染的	L35
innovation /ˌɪnə'veɪʃ(ə)n/ *n.* 创新	L18
inspire /ɪn'spaɪə(r)/ *v.* 赋予灵感	L18
install /ɪn'stɔːl/ *v.* 安装	L34
instruction /ɪn'strʌkʃ(ə)n/ *n.* 指令	L6
integrated /'ɪntɪgreɪtɪd/ *adj.* 集成的，综合的，完整的	L2
integration /ˌɪntɪ'greɪʃ(ə)n/ *n.* 集成；一体化	L14

Intel 英特尔公司	L7
intelligent /ɪnˈtelɪdʒənt/ adj. 智能的	L2
intend /ɪnˈtend/ v. 想要；打算	L34
interchange /ˈɪntətʃeɪndʒ/ n. 互换	L26
interface /ˈɪntəfeɪs/ n. 界面，接口	L14
interfere /ˌɪntəˈfɪə(r)/ v. 干扰	L34
international /ˌɪntəˈnæʃ(ə)nəl/ adj. 国际的	L21
Internet of Things technology 物联网技术	L2、L22
invent /ɪnˈvent/ vt. 发明；创造	L1
inventor /ɪnˈventə(r)/ n. 发明者	L13
irrigation /ˌɪrɪˈɡeɪʃ(ə)n/ n. 灌溉	L23
keep a habit of 保持……习惯	L34
keyboard /ˈkiːbɔːd/ n. 键盘	L5
Kunpeng 鲲鹏	L7
laboratory /ləˈbɒrətri/ n. 实验室	L13
laptop /ˈlæptɒp/ n. 笔记本电脑	L1
launch /lɔːntʃ/ v. 推出（新产品）	L1
layout /ˈleɪaʊt/ n. 布局	L29
lightweight /ˈlaɪtweɪt/ adj. 薄型的	L18
link /lɪŋk/ n. 链接	L29
local /ˈləʊk(ə)l/ adj. 局部的，本地的	L29
logo /ˈləʊɡəʊ/ n. 标志	L29
LOONGSON 龙芯	L7
loudspeaker /laʊdˈspiːkə(r)/ n. 音箱	L25
macro /ˈmækrəʊ/ n. 宏	L35
mainboard /meɪnbɔːd/ n. 主板	L5
maintain /meɪnˈteɪn/ v. 维持	L10
maintenance /ˈmeɪntənəns/ n. 维护	L38
management /ˈmænɪdʒmənt/ n. 管理	L10
manager /ˈmænɪdʒə(r)/ n 管理器	L11
manufacturer /ˌmænjuˈfæktʃərə(r)/ n. 制造商	L7
mass media 大众传媒	L21
medium /ˈmiːdiəm/ n. 媒体	L25

memory /ˈmeməri/ n. 存储器	L3
method /ˈmeθəd/ n. 方法	L10
microphone /ˈmaɪkrəfəʊn/ n. 麦克风	L25
microprocessor /ˌmaɪkrəʊˈprəʊsesə(r)/ n. 微处理器	L1
million /ˈmɪljən/ n. 百万	L22
miniature /ˈmɪnɪtʃə(r)/ adj. 微型的	L2
mixed-signal /mɪkst ˈsɪgnəl/ adj. 混合信号的	L7
Mixed-Signal ICs 混合信号芯片	L7
modernize /ˈmɒdənaɪz/ v. 使现代化	L23
monitor /ˈmɒnɪtə(r)/ v. 监视；n. 显示器	L23
motion /ˈməʊʃn/ n. 移动，运动	L26
mouse /maʊs/ n. 鼠标［器］	L6
multi-core /ˈmʌlti ˈkɔː(r)/ adj. 多核心的	L6
multimedia /ˌmʌltiˈmiːdiə/ n. 多媒体	L2、L22
multipurpose /ˌmʌltiˈpɜːpəs/ adj. 多用途的	L37
navigation /ˌnævɪˈgeɪʃ(ə)n/ n. 导航	L22
networking /ˈnetwɜːkɪŋ/ n. 网络化	L2
oil /ɔɪl/ n. 油；石油	L37
on-demand 按需	L26
operate /ˈɒpəreɪt/ v. 操作	L2
operating system 操作系统	L10
option /ˈɒpʃ(ə)n/ n. 选择，选项	L37
output /ˈaʊtpʊt/ n. 输出	L2
own /əʊn/ adj. 自己的	L30
palm /pɑːm/ n. 手掌	L1
path /pɑːθ/ n. 路径	L11
pathway /ˈpɑːθweɪ/ n. 路径	L19
peer /pɪə(r)/ n. 同龄人	L18
perfect /ˈpɜːfɪkt/ adj. 完美的	L30
perform /pəˈfɔːm/ v. 执行，履行	L9
period /ˈpɪəriəd/ n. 句点	L19
personal /ˈpɜːsən(ə)l/ adj. 个人的	L1
Phytium 飞腾	L7

pixel /ˈpɪksl/ n. 像素	L37
plastic /ˈplæstɪk/ adj. 塑料的	L37
platform /ˈplætfɔːm/ n. 平台	L18
play an important role 起着重要作用	L26
pop-up menu 快捷菜单	L11
portable /ˈpɔːtəb(ə)l/ adj. 轻便的，便携式的	L3
portal /ˈpɔːtl/ n. 门户网站	L21
power /ˈpaʊə(r)/ n. 电力	L5
powerful /ˈpaʊəf(ə)l/ adj. 强大的	L21
presentation /ˌprez(ə)nˈteɪʃ(ə)n/ n. 演示文稿	L17
press /pres/ v. 按下	L11
print /prɪnt/ v. 打印	L19
printer /ˈprɪntə(r)/ n. 打印机	L5
prize /praɪz/ n. 奖赏	L34
process /ˈprəʊses/ v. 加工；处理	L2、L17
processor /ˈprəʊsesə(r)/ n. 处理器；处理程序	L10
product /ˈprɒdʌkt/ n. 产品	L7
professional /prəˈfeʃən(ə)l/ n. 专业人士	L18、L33
programmable /prəʊˈɡræməbl/ adj. 可编程序的	L27
programmer /ˈprəʊɡræmə(r)/ n. 程序员	L13
projector /prəˈdʒektə(r)/ n. 放映机，投影仪	L25
protect /prəˈtekt/ v. 保护	L34
protocol /ˈprəʊtəkɒl/ n. 协议	L14
provide /prəˈvaɪd/ v. 提供	L21
publish /ˈpʌblɪʃ/ n. 发布	L29
random /ˈrændəm/ adj. 随机的	L6
realize /ˈriːəlaɪz/ v. 实现	L23
reboot /ˌriːˈbuːt/ v. 重新启动	L14
recover /rɪˈkʌvə(r)/ v. 恢复	L19
refer to 指的是；提及	L6
register /ˈredʒɪstə(r)/ vt. 注册，登记	L30
registration /ˌredʒɪˈstreɪʃ(ə)n/ n. 注册，登记	L30
registry /ˈredʒɪstri/ n. 注册机构	L30

单词	课次
reinstall /ˌriːɪnˈstɔːl/ v. 重新安装	L37
related /rɪˈleɪtɪd/ adj. 相关的	L14
remote /rɪˈməʊt/ adj. 远程的	L29
require /rɪˈkwaɪə(r)/ vt. 要求	L30
requirement /rɪˈkwaɪəmənt/ n. 要求	L18
rescue /ˈreskjuː/ v. 挽救	L14
research /rɪˈsɜːtʃ/ v. 研究	L33
resource /rɪˈsɔːs/ n. 资源	L21
retail /ˈriːteɪl/ n. 零售	L22
revolution /ˌrevəˈluːʃn/ n. 改革	L2
revolutionary /ˌrevəˈluːʃənəri/ adj. 革命性的	L14
safe /seɪf/ adj. 安全的	L15
satellite /ˈsætəlaɪt/ n. 人造卫星	L22
save /seɪv/ v. 节省；保存	L19
scale /skeɪl/ n. 规模	L2
scandisk /ˈskæn dɪsk/ n. 磁盘检查工具	L38
screen /skriːn/ n. 屏幕	L3
search /sɜːtʃ/ v. 搜索	L21
seasonal /ˈsiːzənəl/ adj. 节令性的	L23
secondary /ˈsekənd(ə)ri/ adj. 辅助的	L6
secure /sɪˈkjʊə(r)/ adj. 安全的	L27
select /sɪˈlekt/ v. 选择；挑选	L6
send /send/ v. 发送	L5
separate /ˈseprət/ v. 分隔	L19
server /ˈsɜːvə(r)/ n. 服务器	L14
signal /ˈsɪgnəl/ n. 信号	L25
site /saɪt/ n. 网站，站点	L29
situation /ˌsɪtʃuˈeɪʃ(ə)n/ n. 形势；情况	L37
size /saɪz/ n. 大小；规模	L3
smart /smɑːt/ adj. 聪明的，智能的	L1
smart irrigation system 智能灌溉系统	L23
smartphone /ˈsmɑːtfəʊn/ n. 智能手机	L1
sound /saʊnd/ n. 声音	L25

单词	课次
special /ˈspeʃ(ə)l/ adj. 特别的；专门的	L30
specific /spəˈsɪfɪk/ adj. 明确的；特殊的	L10
spray /spreɪ/ v. 喷洒	L37
spread /spred/ v. 传播	L34
spreadsheet /ˈspredʃiːt/ n. 电子表格	L17
spyware /ˈspaɪweə(r)/ n. 间谍软件	L33
stable /ˈsteɪb(ə)l/ adj. 稳定的	L15
stand /stænd/ v. 忍受	L39
stand for 代表	L25
standard /ˈstændəd/ adj. 标准的	L14
static /ˈstætɪk/ adj. 静态的	L29
statistic /stəˈtɪstɪk/ n. 统计数字	L22
storage /ˈstɔːrɪdʒ/ n. 存储器，存储	L6
store /stɔː(r)/ v. 存储	L5
streaming /ˈstriːmɪŋ/ adj. 流式	L26
succeed /səkˈsiːd/ v. 成功	L13
successfully /səkˈsesfəli/ adv. 成功地	L33
suggest /səˈdʒest/ v. 建议	L11
suite /swiːt/ n. 套装软件	L18
Sunway TaihuLight 神威·太湖之光（中国超级计算机）	L7
SUNWAY 神威	L7
supercomputer /ˈsuːpəkəmpjuːtə(r)/ n. 超级计算机	L2
supply /səˈplaɪ/ n. 供应	L5
support /səˈpɔːt/ v. 支持	L9
surround /səˈraʊnd/ v. 围绕，包围	L37
SW26010 申威 26010（一款中国自主设计、制造的高性能多线程处理器）	L7
switch /swɪtʃ/ n. 开关，转换；v. 转变；（使）改变	L38
system /ˈsɪstəm/ n. 系统	L9
tablet /ˈtæblət/ n. 平板电脑	L1
take the place of 代替，取代	L23
task /tɑːsk/ n. 任务	L10
technology /tekˈnɒlədʒi/ n. 技术	L2

词汇	课次
telephony /təˈlefəni/ n. 电话（学）	L25
template /ˈempleɪt/ n. 样板	L17
terminal /ˈtɜːmɪnəl/ n. 终端	L22
text /tekst/ n. 文本	L26
thanks to 幸亏；归因于	L23
the Intelligent Earth 智能地球	L2
the Internet of Things 物联网	L2
tip /tɪp/ n. 小窍门，提示	L30
tool /tuːl/ n. 工具	L10
transformation /ˌtrænsfəˈmeɪʃən/ n. 转型	L22
transistor /trænˈzɪstə(r)/ n. [电子] 晶体管	L2
translation /trænzˈleɪʃ(ə)n/ n. 翻译	L10
transportation /ˌtrænspɔːˈteɪʃən/ n. 运输	L22
tray /treɪ/ n. 盘，碟，托盘	L38
trust /trʌst/ v. 相信，信任	L34
tube /tjuːb/ n. 管子	L2
turnover /ˈtɜːnəʊvə(r)/ n. 年营业额	L22
type /taɪp/ v. 打字；输入	L5
typical /ˈtɪpɪk(ə)l/ adj. 典型的	L2
unique /juˈniːk/ adj. 独特的	L30
unit /ˈjuːnɪt/ n. 部件；组件	L6
Universal Plug and Play 通用即插即用	L14
unknowingly /ʌnˈnəʊɪŋli/ adv. 不知不觉地	L33
upgrade /ʌpˈgreɪd/ v. 升级	L34
upload /ʌpˈləʊd/ vt. 上传	L29
user interface 用户界面	L14
vacuum /ˈvækjuːm/ adj. 真空的	L2
vehicle /ˈviːəkl/ n. 交通工具	L22
version /ˈvɜːʃ(ə)n/ n. 版本	L14
video /ˈvɪdiəʊ/ n. 录像；视频	L5
vinegar /ˈvɪnɪgə(r)/ n. 醋	L37
virus /ˈvaɪrəs/ n. 病毒	L9
widely /ˈwaɪdli/ adv. 广泛地	L14

Wi-Fi (wireless fidelity) 基于 IEEE 802.11b 标准的无线局域网　　L2
wipe /waɪp/ v. 揩，擦　　L37
Zhaoxin 兆芯　　L7

郑重声明

高等教育出版社依法对本书享有专有出版权。任何未经许可的复制、销售行为均违反《中华人民共和国著作权法》，其行为人将承担相应的民事责任和行政责任；构成犯罪的，将被依法追究刑事责任。为了维护市场秩序，保护读者的合法权益，避免读者误用盗版书造成不良后果，我社将配合行政执法部门和司法机关对违法犯罪的单位和个人进行严厉打击。社会各界人士如发现上述侵权行为，希望及时举报，我社将奖励举报有功人员。

反盗版举报电话　（010）58581999　58582371
反盗版举报邮箱　dd@hep.com.cn
通信地址　北京市西城区德外大街4号
　　　　　　高等教育出版社法律事务部
邮政编码　100120

读者意见反馈

为收集对教材的意见建议，进一步完善教材编写并做好服务工作，读者可将对本教材的意见建议通过如下渠道反馈至我社。

咨询电话　400-810-0598
反馈邮箱　zz_dzyj@pub.hep.cn
通信地址　北京市朝阳区惠新东街4号富盛大厦1座
　　　　　　高等教育出版社总编辑办公室
邮政编码　100029

防伪查询说明

用户购书后刮开封底防伪涂层，使用手机微信等软件扫描二维码，会跳转至防伪查询网页，获得所购图书详细信息。

防伪客服电话　（010）58582300

学习卡账号使用说明

一、注册/登录

访问 http://abook.hep.com.cn/sve，点击"注册"，在注册页面输入用户名、密码及常用的邮箱进行注册。已注册的用户直接输入用户名和密码登录即可进入"我的课程"页面。

二、课程绑定

点击"我的课程"页面右上方"绑定课程"，在"明码"框中正确输入教材封底防伪标签上的20位数字，点击"确定"完成课程绑定。

三、访问课程

在"正在学习"列表中选择已绑定的课程，点击"进入课程"即可浏览或下载与本书配套的课程资源。刚绑定的课程请在"申请学习"列表中选择相应课程并点击"进入课程"。

如有账号问题，请发邮件至：4a_admin_zz@pub.hep.cn。